CO-AMO-429

Defending Liberty
Pursuing Justice

PUBLISHING

The Practitioner's Guide to Biometrics

William Sloan Coats
Amy Bagdasarian
Tarek Helou
and Taryn Lam

ABA SECTION OF
SCIENCE & TECHNOLOGY LAW

Defending Liberty
Pursuing Justice

11 10 09 08 07   5 4 3 2 1

Cataloging-in-Publication data is on file with the Library of Congress.

Biometrics / William Sloan Coats, editor

Discounts are available for books ordered in bulk. Special consideration is given to state bars, CLE programs, and other bar-related organizations. Inquire at Book Publishing, ABA Publishing, American Bar Association, 321 North Clark Street, Chicago, Illinois 60610.

www.ababooks.org

Contents

Chapter 2
Rethinking Data Protection Regimes to Enable Global Tracking and Prosecution of Terrorists 19

William Sloan Coats, Vickie L. Feeman, and Tarek J. Helou

About the Editors

Amy Bagdasarian is an associate in the Palo Alto office of White & Case LLP. Her practice focuses on intellectual property litigation, including patent, trade secret, and copyright litigation. Ms. Bagdasarian has assisted in the representation of diverse high-technology clients in software, video imaging, portable electronics, semiconductor chip technology, and biomedical technology, among other industries.

Ms. Bagdasarian received her bachelor of arts degree from UCLA and her juris doctor degree from Santa Clara University. She is a member of the California State Bar, the United States District Court for the Northern District of California, and the United States Court of Appeals for the Ninth Circuit.

William Sloan Coats is the executive partner in charge of the Palo Alto office of White & Case LLP. As an intellectual property attorney, Mr. Coats focuses his practice on cases involving software copyrights, patents, trademarks, and trade secret disputes for the software, electronics, and movie industries and bankruptcy issues. He represents leading business, computer, and entertainment hardware and software companies in complex intellectual property matters.

Mr. Coats has held various leadership positions within the American Bar Association's technology-related sections and divisions. He is currently the chair of the ABA Task Force on Biometrics and chair-elect of the Science and Technology Law Section and was a past chair of the Computer Law Division. Mr. Coats is also a member of the Delegation to the United Nations Commission on International Trade Law Working Group on Electronic Commerce.

Throughout his career, Mr. Coats has given many speeches and presentations and has published numerous articles on intellectual property issues in the computer, entertainment, and music industries. Most recently, Mr. Coats gave a presentation titled "Investor Liability After *Grokster*" at the Fourth Annual Rocky Mountain Intellectual Property & Technology Institute Conference. He also gave a presentation on biometric technology at the Fifth Annual Global Privacy Symposium.

Mr. Coats is a member of the California State Bar, the U.S District Courts for the Northern, Central, Eastern, and Southern Districts of California, the U.S. Courts of Appeals for the Ninth Circuit, and the Federal Circuit and the U.S. Supreme Court.

Tarek Helou is an associate in Orrick, Herrington & Sutcliffe LLP's San Francisco office. He is the chairman of the ABA's Committee on Biometrics.

Mr. Helou's practice focuses on white-collar criminal investigations, securities enforcement actions, and intellectual property litigation. He spoke on a panel at the ABA's 2005 Summer Intellectual Property Law Conference. He earned a B.A. from Johns Hopkins University and a J.D. from New York University School of Law.

Taryn Lam is an associate in the Palo Alto office of White & Case LLP. Her practice focuses on intellectual property litigation, including patent, trade secret, and copyright litigation. Ms. Lam has assisted in the representation of high-technology clients in state, federal, and foreign courts.

Ms. Lam received her bachelor of arts degree from Pomona College and her juris doctor degree from the University of California, Berkeley. During law school, she served as an associate editor of the *Asian Law Journal*. She is a member of the California State Bar and the U.S. District Court for the Northern District of California. Ms. Lam is currently vice chair of the American Bar Association's Committee on Biometrics.

Chapter I

Introduction

Tarek J. Helou*

I. Introduction

The use of biometric identifiers had been increasing before members of al-Qaeda hijacked four airliners on September 11, 2001. Government agencies had already required biometric identifiers to control access to some secure buildings and areas. Private companies had started to use biometric identifiers to facilitate retail payments. September 11 led to an increase in calls for the use of biometrics. Most planned new uses have sought to address security shortcomings.

In the near future, you will pay for things, go through airport security, and log onto your computer simply by scanning your iris, retina, or fingerprint. Biometrics are the most accurate form of identifiers and, when used properly, can greatly simplify life. However, biometrics raise new questions about personal privacy, surveillance, and the effects of government and corporate databases that register and hold fingerprint data and other biometric information. Despite these concerns, advocates in the government and private sector claim that the use of biometrics will enhance privacy and reduce identity theft by decreasing reliance on credit cards, Social Security numbers, and passwords, all of which can be lost or stolen easily. Biometrics also raise novel questions of intellectual property law, including who will own the copyrights to data related to or derived from your biometrics and who will have the right to use that data.

Like all technological advancements, biometrics must be used carefully. However, like all technological advancements, biometrics cannot be prohib-

---

* Tarek J. Helou is an associate at Orrick, Herrington & Sutcliffe, LLP, San Francisco.

1

ited from flourishing because of concerns over misuse. Society must adapt to technology because failing to do so is impossible and impedes the advancement of civilization.

A. What Are Biometrics?

Biometrics represent the measurement of any physiological characteristic or personal trait that is distinctive to an individual or a behavioral characteristic.[1] Colloquially, it has come to mean the measurement and matching of physiological characteristics for purposes of identification or verification. Physiological characteristics are unique identifiers because no two people—not even identical twins—have identical biometric measurements. In this sense, biometrics are more accurate than other forms of identification, even DNA testing.[2]

In practice, however, biometric data collection relies upon the creation of a template based on a person's unique characteristics. These characteristics include physical features, such as fingerprints, iris scans, and voice scans. They also include behavioral features, such as gait and handwriting. Therefore, because biometric identifiers use a template, they are *highly reliable but neither perfect nor unique.*

As opposed to being a consistently replicable string of data, such as a Social Security number, biometric samples differ with each recording. For example, the same fingerprint will generate a slightly different sample every time it is recorded. The differences are attributable to several factors, including dissimilar finger placement, poorly maintained collection devices, and even changes in weather conditions, such as humidity or temperature. Some biometrics, like fingerprints, retinal patterns, and iris patterns, are relatively stable and change only through time, injury, or disease. Others, such as facial and voice patterns, are inherently unstable and change frequently. Thus, they are more prone to disguise, manipulation, and incorrect readings.

To deal with these issues, biometric collection devices create algorithms based on user biometrics. A user's stored biometric and the biometric he or she presents to a biometric system would appear different because a small percentage of the biometric data changes with each placement in a biometric scanner. Thus, the biometric-based systems create algorithms that approximate the user's biometric. A biometric algorithm that was a unique identifier would never allow the wrong person to pass for someone (a false acceptance), but would also result in many instances in which an individual's biometric was not matched to the stored biometric (false rejections). By contrast, a system using a biometric algorithm that permits a match with decreased sensitivity will increase false acceptances but decreases false rejections. The use of two biometrics, although more costly and time-consuming, reduces false acceptances and false rejections.

B. Biometrics Are Not a Silver Bullet to Combat Terrorism or Identity Theft

Many people view the use of biometrics as a panacea to future terrorist attacks. This belief caused biometrics to leap to the forefront of national and international debates on security after September 11. This view persists because the use of biometric identifiers is often seen as a foolproof way to identify an individual or verify his or her identity. However, no method of verification or identification is foolproof and *no single technology or system—or group of them—can guarantee security*. Although biometric identifiers are not infallible, they can increase security significantly and they are the most accurate available form of identification.

Biometrics are not perfect and never will be. However, that does not mean that we should not implement better systems. The goal with biometrics, as with any other identification system, is improvement, not perfection.

C. A Description of the Different Types of Biometrics

Physiological biometrics, which are more common and generally more accurate than behavioral biometrics, include fingerprint scans, iris scans, retina scans, hand scans, and facial scans. Behavioral biometrics, which incorporate time and data based on user action, include voice recognition programs, keystroke recognition, and signature recognition. Biometrics can be as simple as recognizing a familiar face or as complicated as a computerized system that analyzes fingerprints and voices.

The most commonly used biometric systems are fingerprints, iris scans, face recognition, and hand geometry. These technologies vary not only in terms of their accuracy but also in the types of applications and facilities for which they are best suited.

1. *Physiological Biometrics*

a. Fingerprints

According to the International Biometric Group, fingerprints represent almost half the market share of biometric technologies. Fingerprints are most common in government settings because they are among the most accurate and least expensive of all biometrics. Fingerprints suffer from a high failure-to-enroll rate because some people cannot generate a clean fingerprint image and accuracy decreases with age. Fingerprint accuracy can be improved when multiple fingers from each individual are enrolled in a system. Fingerprint readers often use poroscopy, measure body heat, and incorporate pulse readers to ensure that the fingerprint offered is from a living person and it is a real finger (as opposed to a fake hand or a latent print). Fingerprints are also the least intrusive and most familiar of all biometrics.

b. Iris Scans

Systems that use iris scans represent approximately 10 percent of the market share of biometric technologies. Iris scans are the most accurate of all biometric technologies. Consequently, many high-security areas use iris scans. Iris scans are more intrusive than fingerprints, as the individual needs to place his or her eye very close to the reader, which also increases the amount of time for each scan. Iris scanners are becoming more powerful and are effective from increased distances. Iris scan systems are more expensive than other types of biometrics scanners.

c. Facial Recognition

Facial recognition systems represent approximately 10 percent of the market share of biometric technologies. One advantage of face recognition systems is that it can be easily confirmed by a system operator, such as a guard, by comparing a picture in a database with the individual's face. Face recognition systems are less intrusive than any biometric but are also less accurate and rely on several external factors, such as camera quality, facial position, facial expression, and other features such as facial hair or sunglasses.

d. Hand Geometry

Systems that use hand geometry represent approximately 10 percent of the market share of biometric technologies. Hand-geometry devices have a higher false acceptance rate than fingerprint scanners. Like fingerprint readers, hand-geometry readers require little user training. Hand-geometry readers are relatively inexpensive.

2. *Behavioral Biometrics*

a. Voice Recognition

Voice or speech patterns represent a small percentage of the market share of biometric technologies. A person says specific words that the system records at enrollment. The system then prompts the person to say one or more of those words when the person is using the system. The system analyzes the speech pattern and determines whether the voice matches the prerecorded version of the words. Voice recognition systems are susceptible to changes in voices created by illness or background noise.

b. Keystroke Recognition

Keystroke recognition analyzes the way an individual types. Users enroll in a system by typing the same word or words several times. The system verifies the user by recognizing the distinctive rhythm a person uses while typing.

c. Signature Recognition

Signature recognition identifies an individual's handwritten signature by scrutinizing the unique way in which a signature is written. Signature verification is different from signature comparison. Signature comparison only examines how the signature looks. Signature verification assesses how the signature was created—instead of addressing the shape of the signature, it looks at changes in the shape, speed, stroke, pressure, and timing that occur during the act of signing.

D. Identification and Verification

The use of biometrics serves two closely related purposes, identification and verification. Identification systems perform "one-to-many" matches and verification systems perform "one-to-one" matches.

1. Identification

Biometric identification is the comparison of a biometric sample offered at the point of transaction to an entire set of stored biometric templates (a one-to-many comparison). The transaction continues if the offered sample matches a stored template. One example of biometric identification is the comparison of a criminal suspect's fingerprint to all fingerprint templates located in the FBI's Integrated Automated Fingerprint Identification System (IAFIS). IAFIS compares a suspect's fingerprint to its catalogue of millions of fingerprints from convicted criminals. Commercial transactions and government benefits systems commonly use biometric identification, too. An identification system that uses a non-unique personal identification number (PIN) (e.g., a telephone number) greatly reduces search times and eviscerates arguments that biometric systems are not fast enough to be used in wide scales (a one-to-few comparison). However, systems with non-unique PINs only work in voluntary systems, such as commercial transactions, and are not used in law enforcement, border control, or intelligence.

2. Verification

Biometric verification is the comparison of a biometric sample offered at the point of transaction to a biometric template that is linked to a token (usually a PIN or access card) that the individual also presents at the point of transaction (a one-to-one comparison). The use of a token in verification systems decreases processing time because a verification system compares the biometric that is presented only to the biometric template that is associated with the presented token. An identification system that uses non-unique PINs combines a tokenless system's flexibility with a verification system's reduced transaction time.

E. Adjustment to Time-Induced Wear on Fingerprints

One complaint about biometric systems is that biometrics, particularly fingerprints, degrade over time. Such arguments must be addressed because fingerprints change over time and they are the most commonly used biometric. A simple solution to this problem is to have the biometric system update the stored template.[3] For example, it could modify the stored template every time it is used, every 100th time it is used, or every two years. This problem will also decrease as biometric readers become more accurate.

F. Biometrics in Use Today

1. National Security and Intelligence

The Department of Defense has started taking and storing fingerprints of "detainees, enemy prisoners of war, civilian internees, and foreigners under U.S. government control who are perceived as national security threats and deemed to require further background checks."[4] This program will be extended to include palm prints, voice sounds, and iris patterns, and the use of several biometrics will enhance the accuracy of this technology. This system will allow the government to identify suspected terrorists who attempt to enter the United States, especially those who use assumed identities or fraudulent documents.

2. Law Enforcement

Fingerprints are often the only evidence investigators uncover that links a suspect to a crime scene or other relevant evidence. Although the reliability of fingerprints as evidence was hardly challenged until recently, criminal defense attorneys have exploited two recent Supreme Court decisions that changed the standards for admissibility of expert witness opinion testimony. The first decision, in *Daubert v. Merrell Dow Pharmaceuticals, Inc.*, 509 U.S. 579 (1993), held that Federal Rule of Evidence 702 forces a trial court to make "a preliminary assessment of whether the reasoning or methodology underlying the testimony is scientifically valid and of whether that reasoning or methodology properly can be applied to the facts in issue."[5] *Daubert* was followed by *Kumho Tire Co., Ltd. v. Carmichael*, 526 U.S. 137 (1999), which made clear that *Daubert*'s holding applied to all expert testimony, not only scientific testimony.[6]

Courts have not accepted defense attorneys' arguments that the methodology behind expert testimony about fingerprint matches is not reliable enough to be admitted as evidence. No appellate court has ruled that testimony about fingerprint matches does not satisfy *Daubert* and *Kumho Tire*.[7] However, some practitioners have taken a different view of the admissibility of fingerprints and have created a virtual cottage industry through articles in law reviews and legal journals attacking court decisions admitting expert testimony about fingerprint matches.[8]

3. *Border Control and Immigration*

The Department of Homeland Security (DHS) has incorporated the collection of biometric identifiers (fingerprint scans and digital photographs) into its procedures for monitoring and regulating certain foreign visitors' entry to and exit from the United States.[9] The program, United States Visitor and Immigrant Status Indicator Technology (US-VISIT), was implemented on January 5, 2004, with entry procedures initially in effect at 115 airports and 14 seaports, and then at the 50 busiest land ports by the end of 2004. As of the end of 2004, the entry procedures applied to the 50 busiest land ports. All entry points will be covered by the end of 2005.

US-VISIT implements congressional mandates[10] requiring that DHS "create an integrated, automated entry-exit system that records the arrival and departure of aliens; that equipment be deployed at all ports of entry to allow verification of aliens' identities and the authentication of their travel documents through comparison of biometric identifiers, and that the entry-exit system record alien arrival and departure information from these biometrically authenticated documents."[11] The initial congressional mandate for an automated entry/exit program was passed in 1996, but the September 11, 2001 terrorist attacks against the United States expedited the formulation and implementation of US-VISIT. The US-VISIT program accomplishes the following general goals: (1) to enhance the security of citizens and visitors; (2) to facilitate legitimate travel and trade; (3) to ensure the integrity of the immigration system; and (4) to protect the privacy of visitors.[12] Fingerprint scans and facial photographs are the technologies of choice to achieve such goals because they are relatively unintrusive and highly effective at establishing identity.

4. *Commercial Applications*

It is not a matter of *if* biometrics will be the cornerstone or a primary part of many companies' future business plans, it is only a matter of *when*. Some retailers have installed fingerprint readers at check-out registers to enable biometric point-of-sale transactions that offer unparalleled convenience for customers. During an initial registration process, the system links a user's fingerprint data to bank accounts and credit card accounts.

The user simply places his finger on the reader and the correct amount is debited from the selected account—there is no need to retrieve a credit card or check. These applications are most common in grocery stores, fast-food restaurants, and cafeterias. Large-scale programs are in use at Blockbuster video and in grocery stores run by Kroger, Piggly Wiggly, and Thriftway. Welsh Valley Middle School in Narberth, Pennsylvania, had a biometric point-of-service payment system in its cafeteria. Students did not have to remember their lunch money, and students who received free and reduced-price lunches were not subject to having their names checked off a list.[13]

Growth in the use of biometric-enabled retail application will be driven largely by the potential for fraud reduction. Check fraud and credit card fraud cost retailers and banks significant amounts of money every year. The use of biometrics will deter most such fraud, but not all.

Consumer convenience is another reason companies will use biometric-based payment systems. Customers will not have to remove their credit cards or checks to make purchases, will no longer sign credit card receipts, will not be required to produce photo identification, and will not risk identity theft by writing their Social Security or driver's license number on checks. Additionally, the use of a non-unique PIN in identification systems reduces transaction times to the same as a regular credit card transaction and does not require a token or PIN that must be kept at all times or remembered.

Biometrics also allow easy tie-ins with loyalty programs, which can be used to introduce customers who are afraid to use biometric technology. Loyalty programs do not require any personal data other than a customer's name and loyalty account number. Many customers use biometric-based loyalty programs and eventually become comfortable enough to enroll credit cards and bank accounts.

5. Government Benefits

Biometric technologies can reduce welfare fraud by discouraging "double-dipping." Double-dipping occurs when people claim benefits under multiple identities. Biometric systems can screen applicants to determine if they are already enrolled under one name and prohibit enrollment in multiple accounts.

6. Physical and Network Access

Biometrics can increase security for access to networks and physical locations. A biometric-based access system can be set up to restrict or allow certain individuals access to different network locations.[14] This would also obviate the need for individuals to remember passwords or carry keys or access cards. Biometrics can also be used to limit access to facilities such as nuclear power plants, sensitive government buildings, office buildings, hospitals, and laboratories that contain hazardous materials such as poisonous chemicals, germs, and nuclear or other radioactive material.

II. Advantages of Biometrics

A. Biometrics Are the Most Accurate Form of Identification

While not perfect, biometric technology may be the most accurate method available today for identifying and monitoring individuals. To begin with, each individual's biometric identifier is unique to him or her. Therefore, biometric-based identification will not cause confusion of one individual for another, even where such individuals have similar names, addresses, etc. Moreover, a person's biometric identifier, be it his finger, iris, or face, cannot be lost or stolen. Finally, the difficulty of successfully forging or imitating another's biometric identifier will deter identify thieves and other wrongdoers who wish to utilize falsified identification papers or cards.

B. Biometrics Are Unique Personal Identifiers

The use of biometrics has increased tremendously over the past few years and will continue to increase. Biometrics will become more common in government identification cards, as means of completing retail transactions, and as accurate methods of restricting access to physical spaces and computer networks. Given the increased security and unparalleled accuracy biometrics offer, it is virtually certain that industry revenues will surpass current forecasts. Additionally, technological improvements have made biometric solutions more viable, less expensive, and more accurate, which will lead to increased acceptance among potential users.[15]

III. What Has Driven Recent Increases in the Use of Biometrics?

A. Demand for Better Identification and Verification

1. National Security

There is no doubt that the use of biometrics can greatly contribute to counter-terrorism measures and improve national security. The tragic events of September 11 revealed the many flaws in the existing intelligence and law enforcement systems and illustrated the dire consequences that could result from the breakdown of these systems. It is very possible that proper use of available biometric technology by government agencies could have enabled these agencies to identify, find, track, and capture the September 11 terrorists before they could have carried out their destructive attacks.[16] Since then, the U.S. government has turned to biometrics to facilitate its efforts to prevent future attacks. In particular, biometrics can substantially increase the ease, accuracy, and effectiveness of data gathering and data sharing among the different government agencies, thus making it possible for officials to detect and incarcerate potential terrorists.

2. Identity Theft

Identity theft is the fastest-growing crime in the United States. Several high-profile incidents in the past six months have increased its visibility.[17] Each victim of identity theft will spend on average nearly $1,500, excluding attorneys' fees, and 600 hours of time to resolve the problems it causes.[18] Increased accumulation of personal data has led to a need to secure that data, as customer and employee databases are prime targets for identity theft. A single vulnerability in a company's databases can grant a thief access to personal data on hundreds of thousands of persons.

The use of biometrics will deter identity theft because biometric payment systems will inhibit anyone from using a credit card or passing a check that belongs to someone else. Additionally, biometric identifiers in passports and driver's licenses will make it much more difficult for an identity thief to travel within the United States or to the United States with a stolen identity document. Furthermore, the use of biometrics will decrease identity theft perpetrated by employees working at a company that stores personal data because the use of biometrically secured network access more effectively limits access and enhances accountability, as discussed below.

B. Increased Supply of Biometrics

Biometric systems are considerably cheaper and more effective than they were even a few years ago. The improved quality in readers has led to an increase in the number of potential uses for biometric-based systems. Improvements in database size and computer speed allow for larger biometric databases. The reduction in price has allowed more organizations to afford to invest in them. Arguments that biometric systems are too expensive to implement are incorrect. As with every new technology, prices start high and decrease significantly over short periods of time. An example is the reduction in the price of Internet access, especially when increased bandwidth is taken into account.

IV. Privacy Concerns

A. Security Gains vs. Privacy Rights

The use of biometrics may raise privacy concerns because the technology facilitates the collection of all kinds of information about a person, including his or her movements, habits, and preferences. Control over one's personal information would be relinquished to the entities that collect and store the information. Furthermore, as a result of the wealth of information made available, the repercussions of identity theft would be especially severe.

However, the privacy concerns raised by the use of biometrics are nothing new. They are the same concerns implicated under the current regime of data collection. Even without the use of biometrics, an individual's personal information can be tracked by the government and private entities through, for example, credit card usage or a Social Security number. Although biometric technology would increase the amount of private information available, it would also increase the difficulty of stealing or exploiting such information. As described in the previous section, biometrics will actually help combat identity theft and misuse of personal information.

B. The Threat of a "Stolen Biometric" Has Been Exaggerated

Some privacy advocates claim that the increased use of biometrics will result in identity theft through a stolen biometric. Such a scenario would happen if a hacker obtained access to a database of biometric identifiers. This alleged danger rests on the faulty premise that the hacking into a database of biometric identifiers would result in the worst kind of identify theft because although your password, PIN, or identity card can be changed or revoked, your biometric cannot.[19] However, this argument ignores what data is actually stored in a biometric-based system. *Biometric-based systems store an algorithm that represents an individual's fingerprint or other biometric. They do not store an image of the actual fingerprint.* By storing identifying information in proprietary algorithms instead of storing an image or copy of the actual biometric (e.g., an algorithm that represents a fingerprint instead of a scan of that fingerprint), the biometrics company provides a level of protection in case their system is hacked or data somehow becomes available.

Additionally, some biometric databases do not store any information related to the biometric. Matching a person's biometric template (e.g., a fingerprint) to one stored on a card can be accomplished without comparing the biometric template to information in a database. In this type of verification system, the organization would not need to keep databases of biometric information because they would only verify that the person presenting an ID card is the person authorized to have that card.

C. The Use of Biometrics as an Identification Tool for Law Enforcement and Intelligence Agencies

For many years, the police and other law enforcement and intelligence agencies have been utilizing biometrics such as fingerprinting and mugshots to identify criminals. Current advancements in biometric technology would simply increase the accuracy and effectiveness of such tools.

D. Database Linkage

1. *Security and Commercial Improvements*

a. Security

Given the increased capabilities of databases and the urgency our government professes in combating terrorism, most gains from the use of biometrics will result from linked databases with different organizations. Linking databases from several government agencies such as the NSA, CIA, and Departments of Defense, Homeland Security, Justice, Transportation, and Treasury with sources from the private sector would provide greater security gains because it would allow investigators to link disparate pieces of information.

A plan by the Defense Department's Defense Advanced Research Projects Agency (DARPA) called the Total Information Awareness (TIA), which was cancelled before it was created, would have linked commercial and government databases and tracked virtually all government and commercial transactions.[20] The development of TIA raised significant privacy concerns because, in essence, TIA would have kept track of everything that people did that was recorded in any way, including where people went, what they did, what they purchased, and with whom they met. The Defense Department claimed that TIA would have "blanked out" any information that could have been used to identify an individual (e.g., names, addresses, telephone numbers, credit card numbers, and Social Security numbers).

Using biometric identifiers could have been the most effective way to balance law enforcement's needs for linked data and accuracy with a guarantee of anonymity, thereby appeasing TIA's proponents and its detractors. TIA would have treated all of those data sources as if they were a single database and recognized behavioral patterns that typified terrorist activity. TIA would have sifted through individuals' activities recorded by those databases and identified people who engaged in behavior that resembled preparation for terrorist attacks, such as the purchase of certain chemicals, repeated travel to particular nations, suspicious financial transactions, or telephone, e-mail, or chat room contact with other suspected or known terrorists. For example, it would have enabled government investigators to know that an individual who wired money to Pakistan regularly had also traveled to Central America and thus could be involved in a plot to smuggle al-Qaeda operatives into the United States through the Mexican border.

b. Commercial

Increased interoperability and database linkage will also improve commercial operation of biometric-related technologies. For example, if one bank uses a biometric reader on its ATMs (Bank A), it will have to allow individuals with ATM cards issued by another bank (Bank B) to bypass the biometric requirement. In this situation, Bank A will not be able to require the use of a biometric, and it will not be able to determine whether someone who withdraws money with an ATM card from Bank B is using a stolen card. It also will probably have to charge higher fees to customers who access its ATM network with a card issued by Bank B. Additionally, although Bank A can require its customers to present a biometric when accessing its network, it will not be able to do so when they use an ATM at Bank B. This, too, will decrease Bank A's ability to prevent fraud.

2. Privacy Concerns

The development of TIA raised significant privacy concerns that resulted in its cancellation because, in essence, it would have allowed the government keep track of everything that people do that is recorded in any way. Such power has never been available to the government, and the potentials for abuse are plentiful and obvious.

E. Cross-Border Privacy Issues

International organizations are also studying the use of biometrics, mainly as a means of securing borders. A United Nations committee recently heard a proposal that seeks to control immigration by requiring universal fingerprinting and registration. The European Union currently requires fingerprinting of all asylum seekers, and the UN proposal cited the likelihood that the EU will extend the requirement to all citizens of its member states. These international databases raise new privacy issues, such as how to apply laws from several countries that apply different legal standards to the transfer of personal information. Such issues have recently begun to pose problems between American and European divisions of multinational corporations.

V. Conclusion

The increased use of biometrics will not create wholesale changes in the way people communicate and do business like networked communications, mainly the Internet, did. However, biometrics have the capability to fundamentally change small portions of most aspects of our society. For example, the increased use of biometrics *could* fundamentally alter immigration and border control. The increased use of biometrics also *could* fundamentally change the way people purchase goods and services. It also *could* become an everyday part of life, from using your fingerprint to log onto your computer at work every day to using your fingerprint to unlock your car door and your voiceprint to start it, to using your iris and your palmprint to unlock the door to your house.

For biometrics to reach their potential, companies that use them must make voluntary efforts to address legitimate privacy concerns. Companies must also refrain from making false promises about the benefits of biometrics. Privacy organizations must ensure that they do not engage in scare tactics to frighten the public into not using biometrics. The public must be accepting of technological advancement and use biometrics, even if cautiously. The government must interfere with biometrics as little as possible, and allow the private sector and the general public to determine what types of industry standards are best suited to increasing the use of biometrics.

Notes

1. A helpful introduction to biometrics and the various types of biometric technology can be found at http://www.ibgweb.com/reports/public/basic_reports.html.

2. Edward P. Richards, *Phenotype v. Genotype: Why Identical Twins Have Different Fingerprints, available at* http://www.forensic-evidence.com/site/ID/ID_Twins.html (fingerprints of identical twins are different, which makes comparisons of fingerprints more accurate than DNA testing because identical twins have indistinguishable DNA).

3. *In Brief: Biometric Imaging Technology Unveiled*, AMER. BANKER, Nov. 21, 2002, at 10.

4. *DoD Aims to Build Database of Prints—Program Targets Possible Terrorists*, MARINE CORPS TIMES, Nov. 1, 2004, at 33, *available at* http://www.biometrics.dod.mil/documents/publicaffairs/marine-corps-Times-Article.pdf).

5. *Daubert*, 509 U.S. at 592-93.

6. *Kumho Tire*, 526 U.S. at 141.

7. *See, e.g.,* United States v. Sanchez-Birrueta, No. 04-30150, 2005 U.S. App. LEXIS 4673 (9th Cir. Mar. 18, 2005); United States v. Mitchell, 365 F.3d 215, 233-46 (3d Cir. 2004); United States v. Collins, 340 F.3d 672, 682-83 (8th Cir. 2003); United States v. Crisp, 324 F.3d 261 (4th Cir. 2003); United States v. Havvard, 260 F.3d 597, 599-601 (7th Cir. 2001).

8. *See, e.g.,* Kristin Romandetti, *Admissibility of Fingerprint Evidence under* Daubert, 45 JURIMETRICS 41, 48 (2005).

9. Additional information related to US-VISIT is available on the DHS Web site (http://www.dhs.gov) and in the *Federal Register,* 8 C.F.R. Parts 214, 215, and 235 (Implementation of the United States Visitor and Immigrant Status Indicator Technology Program).

10. US-VISIT is authorized by the following statutes: (1) Section 2(a) of the Immigration and Naturalization Service Data Management Improvement Act of 2000 (DMIA), Pub. L.

106–215; (2) Section 205 of the Visa Waiver Permanent Program Act of 2000 (VWPPA), Pub. L. 106–396; (3) Section 414 of the Uniting and Strengthening America by Providing Appropriate Tools Required to Intercept and Obstruct Terrorism Act of 2001 (USA PA-TRIOT Act), Pub. L. 107–56; and (4) Section 302 of the Enhanced Border Security and Visa Entry Reform Act of 2002 (Border Security Act), Pub. L. 107–73.

11. Implementation of the United States Visitor and Immigrant Status Indicator Technology Program ("US–VISIT"); Biometric Requirements, 69 Fed. Reg. 468 (Jan. 5, 2004) (to be codified at 8 C.F.R. pts. 214, 215, and 235).

12. *Id.*

13. *Fingerprints Pay for School Lunch*, CBS News, Jan. 24, 2001, http://www.cbsnews.com/stories/2001/01/24/national/main266789.shtml.

14. The use of biometrics or other security devices to control access to specific files or directories is called "logical access." A biometric reader can be attached to a computer to accomplish this.

15. Valerie Malmsten, *Eye Scans—Authentication With Biometrics*, SANS Institute (Nov. 11, 2000), *available at* http://rr.sans.org/authentic/authentic_list.php; Mark Bruno, *That's My Finger*, U.S. Banker (February 2001), *available at* http://www.us-banker.com/usb/articles/usbfeb01-9.shtml.

16. *See, e.g.,* Walter Pincus, *CIA Failed to Share Intelligence on Hijacker; Data Could Have Been Used to Deny Visa*, Washington Post, June 3, 2002, at A1.

17. ChoicePoint, Lexis-Nexis, PayMaxx, Bank of America, Designer Shoe Warehouse, California State University at Chico, Boston College, T-Mobile, and George Mason University all lost control of sensitive personal data from millions of people. Each of those incidents involved stored data, none involved online transactions.

18. *See* Toby J.F. Bishop & John Warren, *Identity Theft: The Next Corporate Liability Wave?*, Corporate Counselor, March 30, 2005.

19. *See, e.g.,* William Abernathy & Lee Tien, *Biometrics: Who's Watching You?*, at http://www.eff.org/Privacy/Surveillance/biometrics/ (last visited May 6, 2005).

20. A detailed description of TIA is *available at* http://www.darpa.mil/darpatech2002/presentations/iao_pdf/slides/poindexteriao.pdf.

Chapter 2

Rethinking Data Protection Regimes to Enable Global Tracking and Prosecution of Terrorists

*William Sloan Coats, Vickie L. Feeman, and Tarek J. Helou**

I. Introduction

In April 2001, a police officer in Broward County, Florida, pulled over Mohammed Atta and cited him for driving without a license.[1] Atta was given a summons to appear in court on May 28, 2001, but he failed to do so, which prompted the presiding judge to issue a warrant for his arrest.[2] Atta was stopped for speeding on July 5, 2001, in a neighboring county. The officer who stopped Atta was unaware that a warrant had been issued for his arrest and gave him a ticket.[3] Atta left the United States two days later. When Atta returned to the United States on July 19, 2001, he breezed through immigration at Atlanta's Hartsfield Airport and received a visa to stay in the United States for several months, even though he was on a federal terrorist "watch list" used to prohibit foreigners from entering the country or delay visa applications pending investigation of suspected terrorists.[4]

* William Sloan Coats is a partner in the Palo Alto office of White & Case LLP. Vickie L. Feeman is a partner in the Silicon Valley office of Orrick, Herrington & Sutcliffe, LLP. Tarek J. Helou is an associate at Orrick, Herrington & Sutcliffe, LLP, San Francisco. The views expressed in this article do not represent those of White & Case LLP or of Orrick, Herrington & Sutcliffe, LLP.

On September 11, 2001, Atta, aided by four other hijackers, piloted American Airlines Flight 11 into the World Trade Center's north tower. Four others crashed a second airplane into the World Trade Center's south tower, a third into the Pentagon, and a fourth hijacked airplane headed for Washington, D.C., crashed in Pennsylvania. All 19 hijackers were members of Osama bin Laden's al-Qaeda organization, which was responsible for several previous attacks on American targets over the past decade.

The results on September 11 were catastrophic. Almost 3,000 Americans died; the twin towers of the World Trade Center, as well as a third building in the complex, collapsed. The Pentagon was severely damaged and all civilian air traffic in the United States was grounded for three days. The economic toll, although impossible to measure precisely, is believed to be close to $100 billion, and life in lower Manhattan took years to return to a pre-September 11 semblance of normalcy.

Of Atta's 18 accomplices, five obtained Social Security numbers by using false identities, and it is suspected that all 19 made up or appropriated other Social Security numbers and used them to obtain false identifications, at least seven in the form of Virginia state identification cards.[5]

So how is it that Atta, a suspected terrorist and fugitive from the law, could escape from two encounters with law enforcement authorities without being arrested, detained, or even questioned? How were Atta and his accomplices able to obtain fraudulent identification cards that enabled them to board the airplanes they hijacked? The attack al-Qaeda perpetrated on September 11, 2001, represents the greatest blunder by law enforcement and intelligence agencies in the history of the United States, surpassing even the failure to predict the Japanese attack on Pearl Harbor.

This attack has forced our nation to wage a war against "terrorism," an enemy that is difficult to define and unlike any our nation has fought in the past. But, as in any war, the enemy must be identified and located before he can be engaged. Because terrorists do not operate like the armed forces of nations the United States has fought in previous wars, they are much more difficult to find. This difficulty will place a premium on methods of identifying, finding, and tracking terrorists as a means of prohibiting them from entering the United States and attacking American interests within our borders or abroad. It will require extensive cooperation and sharing of data among American intelligence and law enforcement agencies and their foreign counterparts throughout the world.

Bin Laden's followers exploited two problems in American intelligence and law enforcement capabilities that permitted them to live in the United States unnoticed for years while emerging from encounters with individuals charged with stopping them: (1) a failure to collect data that enables adequate tracking of individuals, and (2) a failure to share relevant data that is collected.

Addressing these two problems, which hinder our ability to track dangerous people, will be critical to our efforts to dismantle terrorist organizations like al-Qaeda by detaining, prosecuting, and neutralizing their members.

Technology exists today, in the form of biometrics, that can provide, when used properly and extensively, vast improvements in tracking suspected terrorists. Two amendments to new European Union legislation—both to provisions in the EU Data Protection Directive that hinder data sharing between nations—will ensure that American intelligence and law enforcement agencies and their European counterparts continue to share valuable data that facilitates tracking of suspected terrorists.

II. Use of Biometrics as a Means of Identifying Terrorists

An intense focus has surrounded biometrics as an invaluable weapon in our nation's war against terrorism because the use of biometrics as an identifier is the most effective way to find and track people, including suspected terrorists. The new and prominent position the field of biometrics has attained in public dialogue comes as a variety of sources—newspapers, magazines, government officials, scientists, and engineers—focus on the security that biometric devices could bring our society through use by law enforcement and intelligence agencies and the advantages we can gain through their use in the war on terrorism.

Even though no government entity had created a computer system to allow the customs agent or police officer who encountered Atta to realize that he was a suspected terrorist with a warrant issued for his arrest, and no government entity had taken steps to deter his accomplices from obtaining fake identification cards, technology exists to implement such systems. The use of biometrics as individual identifiers can prohibit the use of false identification documents and quickly link data that sits in numerous databases run by different intelligence agencies, law enforcement agencies, and private companies—including information about terrorist watch lists and arrest warrants. Linking data in this manner would have enabled law enforcement agents who detained Atta in the months preceding the September 11 attacks to realize that they should have detained and interrogated him, or at least questioned him extensively.

A. Biometrics Are the Most Accurate Form of Identification

Biometrics are the most accurate method of identification of individuals. Biometrics are commonly used in the form of algorithms that represent physical characteristics, such as fingerprints, iris scans, retina scans, facial scans, palm scans, or hand geometry measurements that are unique to individuals.[6] Unlike other forms of identification, such as passports and driver's licenses, they cannot be doctored or fabricated, and are not susceptible to honest misidentification of individuals with identical or even similar names or other personal information, such as addresses or phone numbers.[7]

B. Biometrics Are Used More Often as Identifiers

The use of biometrics will increase tremendously in the next few years as they become commonplace in forms of government identification, as means of completing retail transactions, and as accurate methods of restricting access to physical spaces and computer networks. Industry revenues totaled nearly $400 million in 2000.[8] By 2005, that number grew to almost $2 billion.[9] Given the increased security and unparalleled accuracy biometrics offer, it is virtually certain that industry revenues will surpass those forecasts.[10] Additionally, technological improvements have made biometric solutions more viable, less expensive, and more accurate, which has led to increased acceptance among potential users.[11] As biometric identification technology continues to improve, public acceptance of its use will increase to even greater levels, which will lead to even higher penetration rates for biometric devices.

C. Biometrics Will Enhance Security

Increases in the use of biometrics in civilian capacities will result in "spin-ons," in which civilian uses of new technology enhance national security. For example, increased use of biometrics as a method of ensuring secure access to facilities such as nuclear power plants, office buildings, hospitals, and laboratories will decrease the likelihood that terrorists will strike them or seize dangerous materials such as poisonous chemicals, germs, and nuclear or other radioactive material from them. Additionally, one way biometrics will transform our lives is often lost amid discussions about fingerprint-embedded driver's licenses and passports containing facial scans: the use of biometrics to authorize everyday financial transactions. People can already pay for goods simply by placing their finger on a fingerprint reader and no longer need to carry credit cards, debit cards, cash, or checks. The linking of biometric characteristics to bank accounts or credit cards and the subsequent increased use of biometrics in transactions will ensure easier and more accurate tracking of individuals and money, which will aid the tracking of terrorists and their finances by decreasing the likelihood that they will dupe intelligence agencies with false identifications or stolen credit cards.

Government use of biometrics will increase, too. Domestically, the Department of Defense has issued over 600,000 biometrically enabled smart cards to military personnel and will issue close to 1 million by the end of 2002.[12] These cards will enable military personnel to fill out paperwork, sign and send encrypted electronic mail, purchase items, and access buildings. Such widespread use by the federal government will encourage state and local governments to increase their use of biometrics. Several states have already discussed adding biometrics to driver's licenses, and one private-sector group has proposed a national biometric identification system that would be voluntary for American citizens but mandatory for all others who travel to or live in the United States.[13]

D. The Use of Biometrics Will Help Fight Terrorism

Increases in the use of biometrics abroad will serve mainly as a means of securing borders and could lead to a wealth of information that will

help American intelligence and law enforcement agencies fight terrorism. For example, Pascal Smet, Belgian Commissioner General for Refugees and Stateless Persons, proposed a universal identification system at
a United Nations meeting in December 2001.[14] His proposal, mainly an
effort to control immigration, would mandate fingerprinting and registration of every person in the world. In advocating a global database that
identified every human, Smet referred to the possibility that the EU might
institute a similar plan that will require fingerprinting and registration of
the citizens of all its member nations.[15]

The EU plan is an outgrowth of its Justice and Home Affairs Council
Meeting of September 20, 2001. At this meeting the Council directed the
European Commission, the EU's policy-making arm, to "examine urgently the relationship between safeguarding internal security and complying with international protection obligations and instruments."[16] The
commission specifically cited a need for pre-entry screening, including
the use of biometric data and cooperation among border guards, intelligence services, immigration agencies, and asylum authorities, and existing requirements that mandate fingerprinting of all asylum seekers.[17] Although the EU plan applies only to asylum seekers and would likely be
extended to all refugees, Smet indicated that storing biometric data for
every European is a possibility because such a system would face "no
technical problems . . . only a question of will and investment."[18]

E. Linked Databases Will Help Fight Terrorism

American efforts to shut down terrorist financing have been slowed
by a failure to share pertinent information among various government
agencies. The compartmentalization that plagues these efforts has led
one senior U.S. official to describe intelligence-sharing efforts between
the CIA and the FBI as "third world."[19]

Most personal data that assists identification and tracking of terrorists—including travel information, financial records, criminal records,
and medical records—is already stored in several databases. However,
these databases are not linked to each other, and searching numerous
databases is time-consuming. Additionally, different storage protocols
can make connecting a suspected terrorist to his travels and receipt of

finances—two common ways to track terrorists—virtually impossible because different databases often store information by different variables, and terrorists often use false or stolen personal information.

Allowing intelligence and law enforcement agencies to easily link information in these databases by attaching biometric identifiers to the data will permit these agencies to track suspected terrorists more rapidly because they will be able to search several databases simultaneously with a suspect's biometric identifier instead of conducting a discrete search for each database. These single searches also will be more accurate because they are based on unique identifiers (biometrics) instead of names, addresses, or identifying numbers, each of which often represents more than one person, and, unlike biometrics, can easily be fabricated or stolen.

F. Biometrics Will Enable Security Gains Without Compromising Privacy Rights

Some privacy advocates have questioned the use of biometrics. They have voiced concerns about the effects on personal privacy of government or corporate databases that register personal information. Concerns of this nature often accompany innovative devices and technologies that have the potential to change society in revolutionary ways.

Concerns expressed over the increased use of biometrics ignore the obvious and extraordinary security gains they will create. Biometric identifiers, while not perfect or foolproof, have three distinct security advantages over traditional methods of monitoring individuals. First, biometric identifiers are unique to each individual and eliminate the possibility of people with identical or similar names being confused for one another. Second, biometric identifiers are part of an individual's body and cannot be lost, stolen, or passed from one person to another, and their use in immigration documents could render passport theft obsolete. Third, biometric identifiers are very difficult to forge and therefore eliminate the possibility of fabricating identification documents, as many of the September 11 hijackers did.

If law enforcement and intelligence agencies had maintained operationally linked biometric identification systems prior to September 11, at least three of the 19 hijackers could have been detained. If the FBI had received all information available to the CIA, "[t]here's no question [it] could have tied all 19 hijackers together."[20]

1. *Biometrics Will Help Combat Identity Theft*

Identity theft is the fastest-growing crime in America. Detractors of biometrics claim that their increased use will make identity theft more common by simply helping to organize personal data in a single spot. Biometrics will, however, reduce identity theft by decreasing reliance on tokens such as credit cards, Social Security numbers, and passwords, which are easily lost or stolen, by connecting personal information to a biometric that is very difficult to forge, such as a fingerprint or iris scan.

Biometrics also will deter identity theft by discouraging those who have access to personal information from misusing it. A person with access to personal data who has to log onto a database that stores the data by using his own biometric will be unable to deny that he looked at restricted data or to claim that someone else retrieved it by using his password without his permission. Preventing identity theft will prohibit terrorists from easily evading tracking by stealing the identities of others.

2. *Racial Profiling Will Be Reduced as Investigators Focus on Suspicious Activity*

One legitimate concern of the ongoing war on terror is the potential for racial profiling of people of Arabic or South Asian descent or others who look like them. The constitutional concerns racial profiling presents are obvious, and even if they are put aside, the law enforcement gains from such practices are dubious at best. For example, most of the terrorist attacks plotted in the United States since September 11 were not perpetrated by someone of Arab descent—not even those acts connected to al-Qaeda—and no degree of racial profiling would have prevented any of them from carrying out their acts.[21] Biometrics will allow investigators to focus instead on reliable indicators of potential terrorist activity by guaranteeing that individuals cannot hide relevant information, such as numerous visits to countries with significant al-Qaeda activity.

3. Biometrics Can Be Used as Identification Tools for Law Enforcement and Intelligence Agencies

Many who complain about the use of biometrics as a potential tool for an expansion of government powers ignore the fact that biometrics have been used for decades by law enforcement agencies in the form of finger-print databases and mug-shot books. Current systems simply increase their effectiveness, which will reduce the risk of arresting and monitoring innocent individuals and groups. Additionally, a great deal of the gains will be accomplished through the monitoring of non-citizens overseas and therefore will not infringe Fourth Amendment rights.

III. The Need for the EU to Amend Its Data Protection Directive

Waging a war on terror requires battling a foe without borders and soldiers who do not wear uniforms. Tracking and detaining known or suspected terrorists, linking them to their sources of funding or unknown accomplices, and ensuring they do not get radioactive materials are imperative. Keeping track of biometric information will help prevent terrorists from entering the United States, because most of the terrorists who have attacked and will attempt future attacks on the United States come from overseas. Indeed, al-Qaeda, the terrorist organization posing the largest threat to the United States, has a presence in more than 50 nations.[22]

An essential task for American intelligence and law enforcement agencies is to improve their ability to identify and track individuals throughout the world, which will also aid in the prosecution of terrorists. Even though a few al-Qaeda operatives are American—Adam Gadahn, John Walker Lindh, Yaser Esam Hamdi, and Jose Padilla have been detained and connected to al-Qaeda—they still traveled repeatedly between the United States and countries with significant al-Qaeda activity. Improved international tracking systems using biometrics would help investigators determine exactly where they went, when they were there, and with whom they traveled and met.

To successfully track and prosecute terrorists, American intelligence and law enforcement agencies must have unconditional access to the personal data of suspected terrorists regardless of where this data is stored. Different nations have different laws governing the transfer of personal data overseas. Many have aided the United States by sharing information that has led to the arrests of suspected al-Qaeda members here and abroad. For example, significant progress has been made through arrests and elimination of al-Qaeda cells throughout the world. All of the operations in these areas occurred in nations with virtually no prohibitions on sending personal data overseas. Freedom from burdensome data transfer restrictions enabled these nations to cooperate with American intelligence and law enforcement agencies by sharing information even in the face of substantial dissent from their citizens. EU members, notably the United Kingdom, Germany, Spain, and France, have also carried out operations that have disrupted al-Qaeda in their own nations.

However, the previously uninterrupted stream of intelligence from EU members is in danger because the European Union Privacy Directive, also known as the EU Data Protection Directive, does not explicitly permit EU nations to transfer personal data to American intelligence and law enforcement agencies.[23] The Directive also impedes the American goal of eliminating terrorist funding because financial institutions are not permitted to take full advantage of the "safe harbor" provisions negotiated between the United States and the EU. The United States must work with the EU to modify the Directive to remedy these deficiencies. Article 26 of the Directive, which exempts all transfers of personal data in the "public interest" from stringent data protection requirements, should be modified to explicitly include the "public interest" of foreign nations, or at least those allied with the EU, such as the United States. The "safe harbor" provisions negotiated between the Department of Commerce and the EU currently permit transfers between some companies that promise to adhere to specified data protection standards and should be modified to extend all their provisions to financial institutions. The United States can then require financial institutions to meet these safe harbor requirements. If the United States does not work with the EU to amend these provisions, the Directive, as currently written, could become a tremendous hindrance to American prosecutions of terrorists.

A. The EU Should Amend Article 26's "Public Interest" Exemption

Article 25 of the Directive permits transfers of "personal" data from EU nations to countries outside the EU "only if . . . the third country in question ensures an adequate level of protection" of the personal data.[24] Personal data is defined loosely as "any information relating to an identified or identifiable natural person . . . who can be identified, directly or indirectly, in particular by reference to an identification number or to one or more factors specific to his physical, physiological, mental, economic, cultural or social identity."[25] This definition encompasses virtually any information related to any individual and undoubtedly includes biometrics. These prohibitions, which apply to all transfers of data—even transfers between different offices of the same company—will greatly hinder the war on terrorism if they are not expanded to explicitly exempt transfers that involve foreign law enforcement and intelligence needs.

The exemption to Article 25, found in Article 26(1), permits derogation of the "adequate" data protection provision for transfers that are "necessary . . . on important public interest grounds[.]"[26] However, this "public interest" derogation does not explicitly include the "public interest" of non-EU nations, and therefore does not necessarily apply to transfers of personal information to American intelligence and law enforcement agencies.

1. Verbal Assurances from EU Nations That They Will Interpret the Exemption to Apply to American "Public Interest" Are Inadequate

Although certain EU officials have stated that the Directive does not apply to national security activities, national security concerns explicitly exempt certain Articles of the Directive, but not Article 25.[27] Additionally, though the Directive makes "a clear exception when disclosure is necessary for security or law enforcement purposes," it is not clear that this exception applies to non-EU governments.[28] These simple assurances from EU governments that they will interpret the Directive to exempt foreign intelligence and law enforcement agencies from the data transfer requirements are insufficient because interpretations that permit transfers to foreign agencies pursuant to Article 26's "public interest" exemption can easily change to the detriment of efforts to prosecute terrorists. Changes of this nature could subject American national security to the whims of changing political sentiment.

Such a clear exemption in the Directive is necessary, especially considering that the Directive only mandates general principles and requires each member nation to pass its own legislation to apply them. This could easily result in a patchwork system of personal data transfers throughout the EU, with some member nations interpreting the "public interest" exception to include the public interest of foreign nations and others supporting a much narrower interpretation that includes only the public interest of EU nations. The absence of clearly defined permission for data transfers to and from non-EU law enforcement and intelligence agencies such as the CIA, NSA, FBI, and intelligence branches of the U.S. Military could encourage EU nations to craft their own legislation in response to the Directive's "public interest" exemption that: (1) refuses all personal data transfers to and from foreign intelligence and law enforcement agencies; (2) permits them to refuse personal data transfers in certain instances, such as death penalty cases, which could include many terrorism prosecutions; or (3) is vague enough to permit a case-by-case analysis of personal data that could stymie American efforts to prosecute terrorists. Any of these scenarios will severely hinder American prosecutions and decrease the deterrent effect of such prosecutions.

2. Political Disagreements Hinder Data Sharing Between EU Members and the United States

Even though EU nations have, for the most part, cooperated with the United States in recent terrorism investigations, rifts have already developed regarding transfers of information related to suspected terrorists. For example, the EU's rejection of the death penalty conflicts directly with America's willingness to use it in prosecutions of accused terrorists.[29] This disagreement has already hindered American prosecutions of terrorists, as France flatly refused to offer any aid or assistance in the trial of Zacarias Moussaoui, a French national arrested on August 17, 2001, who faced the death penalty after being indicted for allegedly plotting with al-Qaeda to serve as the 20th hijacker on September 11.[30] Germany has announced that it will severely limit evidence it supplies to the United States in the *Moussaoui* case because of the likelihood that he will receive a death sentence if convicted.[31] France and Germany, which are less reliant on American aid and political support than other EU nations and therefore are less likely to be persuaded to cooperate in prosecutions, are the EU nations with the most suspected al-Qaeda activity. Although explicitly permitting transfers of personal data would not require EU nations to share evidence with the United States, such a change would give political cover to leaders of those nations that wish to assist American investigators by providing information that the Directive might restrict.[32]

Other political disagreements between EU nations and the United States could easily turn information sharing into a political football. Likely examples include clashes over a recent shift in U.S. policy toward preemptive attacks, including the use of nuclear weapons to fight terrorists and nations that support them, unilateral American military action against Iraq, and the U.S.'s refusal to grant prisoner of war status to al-Qaeda and Taliban captives and practice of detaining them indefinitely.

B. The EU Should Expand Article 25's "Safe Harbor" Exemption Benefits to Apply to Financial Institutions

The Directive obstructs American efforts to enlist the private sector in fighting terrorism. The U.S. Department of Commerce has negotiated a "safe harbor" provision that permits private companies that agree to adhere to the requirements of the Directive to transfer data as if they were in a nation deemed to have adequate data protection. However, these protections do not apply to data transfers between financial institutions.[33] Modifications permitting transfers between financial institutions are necessary to track and halt funding of terrorist organizations.

1. Financial Institution Biometric Databases Will Be the Largest in the Private Sector

As discussed above, the use of biometrics as identifiers will become customary, and financial institutions, which will see the greatest increase in the use of biometrics, will create extensive, robust biometric databases. Customers will make purchases without the use of any tokens (credit cards, bank cards, checks, or cash) and will register their biometrics as a means of simplifying their lives.[34] Financial institutions will probably follow suit by embedding biometrics in credit cards, ATM cards, and debit cards. As the number of individuals who choose to register their biometrics with financial institutions grows, the value of this information to intelligence agencies will increase proportionately. Because the use of biometric identifiers is superior to any other approach to tracking individuals, and financial institutions will control databases that store biometric information for large numbers of people, it would be prudent, when possible, to allow them to share information to expand their databases.

2. Intelligence and Law Enforcement Agencies Will Increasingly Rely on Private-Sector Data Collection

Financial institutions should be permitted to transfer information according to the "safe harbor" provisions because intelligence and law enforcement agencies will increasingly use data collected by private-sector companies, such as financial institutions. Allowing financial institutions to transfer data will increase the amount of information in each database and will give the government access to data of a higher quality than government-collected data.

The increase in government use of data collected by private-sector companies is a natural outgrowth of the increase in private-sector data collection. For example, Carnivore, the FBI's much-discussed "Internet wiretap" that reads suspects' e-mails and monitors their Web browsing, was used only twice between January 1, 2001, and mid-August 2001, because Internet service providers were able to satisfy the government's needs in every other instance.[35] The bolstering of reporting requirements for financial transfers in response to September 11 will allow the government to increase its reliance on data collected by the private sector because allowing these institutions to share personal data, such as biometric samples, will increase the probability that a suspected terrorist's biometric information is located in multiple databases instead of just one. This in turn will facilitate government searches for data related to suspected terrorists.

Requiring a biometric identifier for overseas transfers, a distinct possibility in the near future, will facilitate the tracking of terrorists' financial support and will require an expansion of the "safe harbor" provisions to apply to all companies. Additionally, biometrics obtained voluntarily by private-sector companies are usually fingerprints, which offer greater accuracy than facial scans, the most common method for surreptitious governmental gathering of biometric samples.

IV. Conclusion

The fight against terrorism is dependent on identifying, finding, and tracking terrorists. Unlike previous wars, in which opponents fielded large armies in predictable locations that were easily discovered, terrorists operate individually or in small groups and are spread throughout the world. Consequently, access to data that confirms where and when they travel, with whom they travel, what they purchase, and where they purchase it will be vital to our nation's survival. The increased use of biometrics as identifying devices will simplify the collection and management of data and improve its accuracy.

In carrying out this war, the United States must rely heavily on its allies to collect important information. Reliance of this kind demands that impediments to fast access to information that identifies terrorists be eliminated. Guarantees that EU nations will interpret the EU's Data Protection Directive to permit transfers of data in instances of an American "public interest" are insufficient because these interpretations will change as political climates change.

Terrorists' sources of funding must be discovered and tracked before they can be eliminated. Financial institutions, the driving force behind tracking terrorists' finances, must be permitted to track this information and, to do so, need access to all of the safe harbor provisions that other companies enjoy.

Notes

1. George Lardner, Jr., *Atta Skillfully Avoided Authorities in U.S.*, Wash. Post, Oct. 20, 2001 at A3.

2. *Id.*

3. *Hijacker Traffic Stop May Be Clue*, AP, January 21, 2002, LEXIS, News Library, AP File.

4. *Id.*; John L. Smith, *Atta's Trail of Terror Passed Through Las Vegas*, Las Vegas Review-Journal, Sept. 20, 2001, at 1A.

> Atta's ability to avoid detection was not unusual for the hijackers: Khalid Almihdhar and Nawaf Alhazmi were not placed on the same terrorist watch list because CIA officials failed to notify their counterparts at the State

Department, FBI, or INS that he was a suspected terrorist until after his last trip into the U.S.

Walter Pincus, *CIA Failed to Share Intelligence on Hijacker; Data Could Have Been Used to Deny Visa*, WASH. POST, June 3, 2002, at A1.

5. Robert O'Harrow Jr. & Jonathan Krim, *National ID Card Gaining Support*, WASH. POST, Dec. 17, 2001, at A1.

6. A helpful introduction to biometrics and the various types of biometric technology can be found at http://www.ibgweb.com/reports/public/basic_reports.html.

7. Confusion over the true identity of the September 11th hijackers was a significant obstacle toward investigating the attack because many hijackers had stolen innocent people's identifications, but also because Saudis typically use four names but their passports only have their first and last names in English, and last names are often tribal names shared by hundreds of thousands of people.

Walter Pincus, *4 Suspects' Identities in Doubt; Confusion Over Hijackers Names Hindering Investigation*, WASH POST, Oct. 6, 2001, at A16.

8. International Biometric Group, *Biometrics Market and Industry Report 2006-2010*, summary *available at* http://www.biometricgroup.com/reports/public/market_report.html.

9. *Id.*

10. Senators Diane Feinstein and John Kyl recently proposed legislation that would require all foreigners entering the United States to possess passports or other identification that contains biometric data. Press Release, Senators Feinstein and Kyl to Introduce Bill to Strengthen Counter-Terrorism Efforts at U.S. Ports of Entry (Oct. 25, 2001), *available at* http://www.senate.gov/~feinstein/releases01/visaso01.htm.

11. Valerie Malmsten, *Eye Scans – Authentication With Biometrics*; SANS Institute (November 11, 2000), *available at* http://rr.sans.org/authentic/authentic_list.php; Mark Bruno, *That's My Finger*, U.S. BANKER, February 2001, *available at* http://www.us-banker.com/usb/articles/usbfeb01-9.shtml.

12. *DoD Common Access Card Information Brief*, http://egov.gov/smartgov/information/dixon-butler_update/sld002.htm (last visited June 13, 2002).

13. Larry Ellison, Opinion-Editorial, *Digital IDs Can Help Prevent Terrorism*, WALL ST. J., Oct. 8, 2001, at A26 (Ellison offered to donate the necessary software but expects to gain revenue from maintenance, support, and upgrades); Interview by Elise Ackerman with Larry Ellison, CEO, Oracle Corp., *Ellison Goes into Detail About National ID*, *available at* http://www.siliconvalley.com/docs/hottopics/attack/ellisn102001.htm.

14. *UN: Refugees Meeting Hears Proposal to Register Every Human in World*, AAP, Dec. 14, 2001, LEXIS, News Library, AAP Newsfeed File.

15. *Id.*

16. Commission Working Document: The Relationship Between Safeguarding Internal Security and Complying with International Protection Obligations and Instruments, COM(2001)743 final at 6.

17. *Id.* at 6, 18. The Commission noted that all asylum seekers of at least 14 years of age in EU nations must register their fingerprints pursuant to the Dublin Convention.

18. AAP, *supra* note 14.

19. Karen DeYoung & Douglas Farah, *Infighting Slows Hunt for Hidden Al Qaeda Assets*, WASH. POST, June 18, 2002, at A1.

20. Pincus, *supra* note 4.

21. The FBI has expressed its virtual certainty that a native-born individual introduced anthrax into the U.S. Mail system in the Fall of 2001, virtually eliminating Arab-Americans from suspicion. The teenager who flew a plane into a Tampa office building on Jan. 5, 2002 after expressing sympathy for al-Qaeda was not of Arabic descent. The Jewish Defense League members arrested on Dec. 11, 2001, for conspiring to bomb the office of Congressman Darrell Issa are not of Arabic descent. Richard Reid, an al-Qaeda member arrested on Dec. 22 for attempting to destroy a plane in flight from Paris to Miami with a "shoe bomb," is not of Arabic descent. Lucas Helder, arrested on May 7, 2002 for setting off several pipe bombs in the Midwest in the Spring of 2002, is not of Arabic descent. Jose Padilla, an al-Qaeda operative arrested on May 8, 2002. at Chicago's O'Hare airport for plotting to release a "dirty bomb" in the U.S., is not of Arabic descent.

22. Robert S. Mueller III, director, Federal Bureau of Investigation, Remarks at the Anti-Defamation League's 24th Annual National Leadership Conference (May 7, 2002) (transcript *available at* http://www.fbi.gov/pressrel/speeches/speech050702.htm).

23. The official name is *The Directive 95/46/EC of the European Parliament and of the Council of 24 October 1995 on the protection of individuals with regard to the processing of personal data and on the free movement of such data.* EU members are Belgium, Denmark, Germany, Greece, Spain, France, Ireland, Italy, Luxembourg, the Netherlands, Austria, Portugal, Finland, Sweden, and the United Kingdom. Candidates for EU membership are Bulgaria, Czech Republic, Estonia, Cyprus, Latvia, Lithuania, Hungary, Malta, Poland, Romania, Slovenia, Slovakia, and Turkey.

24. Council Directive 95/46/EC, 1995 O.J. (L 281) 31, Art. 25 cl. 1. Only three countries—Canada, Hungary, and Switzerland—ensure a level of protection that the EU has deemed "adequate."

25. Council Directive 95/46/EC, 1995 O.J. (L 281) 31, Art. 2 sec. (a).

26. Council Directive 95/46/EC, 1995 O.J. (L 281) 31, Art. 26 cl. 1(4).

27. John F. Mogg, Director General, European Commission, DG Internal Market, Remarks at the "Privacy—Human Right" International Conference (Sept. 26, 2001) (transcript *available at* http://www.paris-conference-2001.org/eng/contribution/mogg_contrib.pdf).

28. *Id.* The Directive went into effect in 1998, and most of its provisions have yet to be tested. See a similar discussion relating to personal data transfers involving the U.S. Army's collection of biometric data in Europe in JOHN D. WOODWARD, JR. ET AL., ARMY BIOMETRIC APPLICATIONS: IDENTIFYING AND ADDRESSING SOCIOCULTURAL CONCERNS, App. C (RAND Corp. 2001), *available at* http://www.rand.org/publications/MR/MR1237/MR1237.appc.pdf.

29. Abolishment of the death penalty is a requirement for admission to the EU. The United States is already seeking the death penalty for accused terrorists Zacarias Moussaoui and Richard Reid, and it is likely that the United States will seek the death penalty in virtually all terrorism prosecutions. Federal prosecutors probably would have sought the death penalty when prosecuting the defendants who bombed the World Trade Center in 1993 but the statute permitting it was not passed until 1994.

30. Peter Finn, *Germany Reluctant to Aid Prosecution of Moussaoui, Concerns About Death Penalty Hinder Cooperation on Evidence*, WASH. POST, June 11, 2002, at A1. Even though France gave the U.S. assistance, it was not used at trial because Moussaoui pleaded guilty.

31. *Germany to Limit Evidence It Provides to U.S.*, WASH. POST, June 12, 2002, at A28.

32. Some EU nations are willing to circumvent opposition to the death penalty when possible, as evidenced by the UK's agreement to hand over to the United States for arrest certain detainees its forces capture in Afghanistan instead of arresting them, which would have led to complications with their extraditions.

33. Nicola Marsden & Vinod Bange, *Data Transfers Via the EU: Navigating the Route to Compliance*, ELECTRONIC BANKING LAW & COMMERCE REPORT, April 2002, at 11.

34. Descriptions of Indivos's patented biometric products, including Pay-By-Touch, its patented tokenless biometric transaction system, can be seen on its Web site at http://www.indivos.com.

35. Dan Eggen, *'Carnivore' Glitches Blamed for FBI Woes*, WASH. POST, May 29, 2002 at A7.

Chapter 3

United States Visitor and Immigrant Status Indicator Technology Program

*Michael R. Hoernlein**

I. Introduction

The ease with which 19 hijackers obtained U.S. visas, entered the United States, and boarded the planes used in the September 11, 2001 terrorist attacks reflected systemic problems within several government agencies.[1] Two systemic weaknesses identified by the National Commission on Terrorist Attacks Upon the United States (the 9/11 Commission) are: "a lack of well-developed counterterrorism measures as a part of border security and an immigration system not able to deliver on its basic commitments, much less support counterterrorism."[2] In addition, according to the 9/11 Commission, no government agency "systematically analyzed terrorists' travel strategies," which were designed to exploit weaknesses in U.S. border security. The 9/11 Commission found that as many as 15 of the hijackers "were potentially vulnerable to interception by border authorities."[3]

---

* Michael R. Hoernlein is an associate with Dewey Ballantine LLP, New York, New York.

The congressional mandate for an automated entry-exit program to regu-late the movement of non-citizens into and out of this country predated the September 11 attacks. However, the attacks expedited the formulation and implementation of what has become the United States Visitor and Immigrant Status Indicator Technology (US-VISIT) program. The attacks also compelled Congress to require that the program incorporate the collection of biometric identifiers.

The Department of Homeland Security (DHS) began the phased imple-mentation of US-VISIT, which currently uses digital fingerprints and photo-graphs to identify people, on January 5, 2004, with entry procedures initially in effect at 115 airports and 14 seaports.[4] As of the end of 2004, the entry procedures applied to the 50 busiest land ports. By the end of 2005, all entry points are to be covered. In addition, a pilot program of automated exit proce-dures began at Baltimore-Washington International Airport and Miami Inter-national Cruise Line Terminal and, by November 2004, included 13 addi-tional exits points.[5]

US-VISIT's use of biometrics addresses some of the problems identified by the 9/11 Commission, but much work remains. This case study will pro-vide a brief overview of the US-VISIT program and address a few salient issues. It is not intended to be a comprehensive critique of the program.

II. Persons Subject to US-VISIT

With certain exceptions, US-VISIT applies to visitors who seek to enter the United States (or who have entered the United States) on non-immigrant visas (i.e., those who reside permanently outside the United States but wish to enter the country on a temporary basis), and, as of September 30, 2004, visi-tors entering under the Visa Waiver Program.[6] The exceptions include:

- Visitors admitted on an A-1, A-2, C-3, G-1, G-2, G-3, G-4, NATO-1, NATO-2, NATO-3, NATO-4, NATO-5, or NATO-6 visa (generally, these are certain foreign government officials and members of their immediate family); Taiwan officials who hold E-1 visas and members of their immediate families who hold E-1 visas;
- Persons under the age of 14 or over the age of 79;
- Classes of visitors held exempt by joint determination of the Secretary of State and the Secretary of Homeland Security; and

- Any individual visitors held exempt by joint determination of the Secretary of State and the Secretary of Homeland Security or the Director of the Central Intelligence Agency.

Because most Canadian nationals do not require a visa to enter the United States, US-VISIT is generally inapplicable to them. However, Canadians who are required to obtain a visa to enter the United States are subject to US-VISIT.

Similarly, most Mexican nationals who travel to the United States are also not subject to US-VISIT. Most Mexican citizens who enter the United States use a B1/B2 visa/Border Crossing Card (BCC) (also known as a "laser visa").[7] According to the Department of Homeland Security, Mexican citizens who use a laser visa "only as a BCC will [initially] not be subject to US-VISIT procedures."[8] However, "[t]his is an interim solution for the land border while the Department explores a long-term solution to record the entry and exit of visitors crossing our land ports of entry."[9] Mexican citizens who use laser visas to travel outside the "border zone" or to stay in the country longer than 30 days "will be processed through US-VISIT at the land border secondary inspection areas."[10]

The first obvious limitation of US-VISIT, therefore, is that it does not apply to most people entering the United States, since a vast majority of the people who enter the United States each year do so by land along the Canadian and Mexican borders.[11] Many of these people are simply not subject to the US-VISIT procedures for the reasons discussed above. In addition, people who enter the United States illegally by avoiding official entry points altogether are off US-VISIT's radar entirely.

III. US-VISIT Procedures

US-VISIT implements congressional mandates[12] requiring (i) that the DHS "create an integrated, automated entry/exit system that records the arrival and departure of aliens"; (ii) "that equipment be deployed at all ports of entry to allow verification of aliens' identities and the authentication of their travel documents through comparison of biometric identifiers"; and (iii) "that the entry/exit system record alien arrival and departure information from these biometrically authenticated documents."[13]

To that end, US-VISIT, among other things, incorporates the collection of biometric identifiers into the visa-application process, the procedures for allowing certain non-citizens entry into the United States, and (only to a limited extent currently) the procedures for the exit of non-citizens from the United States. Because of extreme time constraints faced in the aftermath of the September 11 attacks, the architects of US-VISIT were forced to integrate and modify the following preexisting data management systems instead of designing the electronic framework for US-VISIT from scratch:[14]

- the Arrival and Departure Information System (ADIS);
- the Passenger Processing Component of the Treasury Enforcement Communications System (which includes the Interagency Border Inspection System (IBIS) and the Advance Passenger Information System (APIS)); and
- the Automated Biometric Identification System (IDENT).

US-VISIT also interfaces with other DHS and non-DHS systems. The result is a complex web of interrelated databases that process, store, and share information.

A. Visa Application

US-VISIT requires that, in addition to providing biographic information as before, an applicant for a visa also submit to fingerprinting (via digital inkless scan of both index fingers[15]) and digital photograph by a State Department official at the visa-issuing post overseas. The fingerprints and photograph are entered into the IDENT system and compared with other information already stored in IDENT. Since 1995 and prior to the September 11 attacks, only a mandatory name check of visa applicants was conducted against a collection of databases called the Consular Lookout and Automated Support System (CLASS), which contains information such as prior visa denials and the existence of federal arrest warrants. Also included within CLASS is TIPOFF, a database containing a watch list of known or suspected terrorists. The biographic and biometric entries are compared against the system of databases before a decision is made about the applicant's admissibility.

B. Entry Procedures

Airlines and cruise lines transmit crew and passenger manifests using the Advance Passenger Information System (APIS). The information is compared with the biographic databases and used to determine if biometric information is available for anyone on the list. Upon arrival at a U.S. entry port, a U.S. Customs and Border Protection Officer collects the visitor's biometric information by again digitally scanning each index finger and taking a digital photograph. The data is used to verify the identity of the visitor and is compared against IDENT and the relevant watch lists. As before, the Customs and Border Protection Officer also inquires about the visitor's stay in the United States. If necessary, the officer will conduct a further investigation based on the results of the verification process. If there are any national security or law enforcement concerns, improper documentation, or any other grounds for preventing the visitor from entering the country or delaying entry, the officer will require further screening of the visitor. Otherwise, the visitor will be admitted.

As the 9/11 Commission Staff makes clear, many of the tactics used by terrorists throughout the 1990s were also used in the planning and execution of the September 11 attacks. The use of biometrics in the entry/exit process makes some of the tactics especially difficult for potential terrorists in the future. For example, "photo-substituted" passports are substantially more difficult to employ with the introduction of digital photograph and fingerprint checks.

C. Exit Procedures

As discussed above, US-VISIT's pilot exit procedures are currently in use at a limited number of exit points. The purpose of the pilot program was to determine the most effective method for monitoring foreign visitors' exit from the country. The following three alternatives are being tested:

- Checkout at a kiosk, which resembles a small automatic teller machine (ATM), within the airport or seaport at which a visitor's travel documents are read, fingers are scanned and a photograph is taken, after which the visitor receives a receipt;

- In addition to the procedures followed in the first alternative, the visitor presents the receipt at the departure gate and provides a single finger scan before boarding; and
- Biometric checkout at the departure gate with a US-VISIT attendant.

The exit procedures are currently being implemented only at the air and sea ports listed previously (no land ports), although the goal is ultimately to institute exit procedures at all air, land and seaport points of entry. In the meantime, this remains a significant limitation of the program. Non-immigrant visa holders (other than those subject to the National Security Entry/Exit Registration System (NSEERS)[16]) need not exit the country through a port at which biometric information is collected. However, the DHS recommends that such visitors retain evidence of departure (such as a passport stamp or plain ticket stub) in case a question arises in the future about whether they overextended their stay.

The 9/11 Commission Staff identified overstaying visas as another important pre-September 11 problem. For example, although the Central Intelligence Agency had provided information about two of the hijackers to border and law enforcement authorities, there was no way to know if the men were still in the country.[17] Until US-VISIT's exit procedures are fully in place, the usefulness of the program in enforcing our immigration laws and in preventing future terrorist attacks is severely limited.

IV. Goals of US-VISIT

The DHS cites the following general goals of the US-VISIT program:

- To enhance the security of citizens and visitors;
- To facilitate legitimate travel and trade;
- To ensure the integrity of the immigration system; and
- To protect the privacy of visitors.[18]

Enhancing security has been a national priority since the September 11 terrorist attacks. According to the DHS, US-VISIT's goal of enhancing security is achieved by (1) using better data and better data access to prevent high-threat aliens or otherwise inadmissible persons from entering the country, (2) using "improved identification of individuals who may be inadmissible to the United States" to reduce the risk of terrorist attacks and illegal immigration, and (3) "improv[ing] cooperation across Federal, State and local agencies through better access to data on foreign nationals."[19] US-VISIT, therefore, at least partially addresses a common criticism of the government's pre-September 11 procedures: lack of information sharing among government agencies.[20] It also makes the State Department a much more important part of counterterrorism than it had been prior to the September 11 attacks.

The focus on enhancing security has raised concerns not only about the residual effects on personal privacy, but also about the creation of additional obstacles (such as time and effort) to legitimate domestic and international travel. US-VISIT's goal of facilitating legitimate travel is achieved by reducing the amount of time required to determine the status (such as inadmissibility) of non-immigrant aliens and increasing the accuracy of such determinations.[21] According to the DHS, the US-VISIT procedures actually make it faster and easier for people to obtain visas. There is also a minimal impact on processing time at entry points. According to the DHS, the incremental increase in processing time due to the biometric component of US-VISIT is 15 seconds.

The September 11 attacks also resulted in increased scrutiny of potential holes in the immigration system. According to the DHS, US-VISIT ensures the integrity of the immigration system by (1) improving the enforcement of immigration laws using better and more complete data, (2) reducing the number of non-immigrant aliens overstaying their visit, and (3) utilizing existing information systems to improve interagency (federal, state, and local intelligence and law enforcement) information sharing. Once again, facilitating the collection and sharing of better information is the primary benefit of US-VISIT.

The more expansive collection of personal information in the form of biometric identifiers implicates personal privacy issues. US-VISIT includes measures to ensure the protection and proper use of private information, as discussed below.

V. Choice of Biometric Identifiers

As discussed, US-VISIT currently involves inkless fingerprint scans and digital photographs. However, the statutes authorizing US-VISIT do not restrict the DHS to these identifiers, so DHS may incorporate the collection of other biometric identifiers "where doing so will improve the border management, national security, and public safety purpose of the entry/exit system."[22]

The DHS's choice to begin the program using fingerprint scans and digital photographs as opposed to other methods of identification is, in part, due to the relative lack of intrusiveness of the collection of these biometric identifiers and their effectiveness at establishing identity.[23] The DHS also notes that these have historically been the preferred identifiers by "the law enforcement communities and the travel industry."[24] The DHS tried to strike a balance between cost and effectiveness while taking into account the burden on those wishing to enter the country.

The Department of Justice, the National Institute of Standards and Technology, and the Department of State issued a report to Congress in January 2003 titled "Use of Technology Standards and Interoperable Databases with Machine-Readable, Tamper-Resistant Travel Documents." The report recommended the use of two-finger scans and a digital photograph for "one-to-one verification" matching, but where matching against a large database ("one-to-many identification" matching), the report recommended the use of 10-finger scans. Since the report, further tests have added support for their recommendation.

VI. How the Data Is Used/Privacy Issues

The data collected from the US-VISIT procedures (which includes biometric identifiers, biographic information, and details of arrival and departure) is stored in relevant DHS databases. The current policy of retaining the information for 100 years is under review.

Because US-VISIT does not apply to U.S. citizens, the program does not implicate the full scope of potential constitutional issues relating to privacy. Nevertheless, there is a legitimate concern regarding the security of the personal information that is collected.

The data is primarily accessible on a need-to-know basis for official purposes by authorized officers from the following agencies: Customs and Border Protection, U.S. Immigration and Customs Enforcement, U.S. Citizen and Immigration Services, as well as consular offices of the Department of State. The information may also be used by law enforcement officers under certain circumstances. According to the DHS, "physical, administrative, technical, administrative and environmental" safeguards are used to protect this data.[25]

On February 23, 2005, the DHS announced the creation of a 20-member Data Privacy and Integrity Advisory Committee to advise the Department of

Homeland Security Secretary and Chief Privacy Officer "on programmatic, policy, operational, and technological issues that affect privacy, data integrity, and data interoperability in DHS programs."[26] The committee's members bring expertise in such areas as "privacy, security, and emerging technology" from careers in business, academia, and the nonprofit sector.[27]

VII. Conclusion

As of September 26, 2005, about 38 million visitors were processed by US-VISIT, and DHS has "taken adverse action against more than 850 criminals and immigration violators."[28] DHS continues to collect data and to develop the systems and procedures for doing so.

Although terrorists continuously plot to exploit our existing weaknesses and to overcome our new security methods, the use of biometric technology by border security agencies undoubtedly makes it harder for terrorists to travel by conventional means. In addition, improved cooperation among government agencies and better integration of both data and screening functions provide greater integrity to our border security. US-VISIT, however, is not fully deployed, and the government has much work to do to complete it.

Notes

1. *See generally* NATIONAL COMMISSION ON TERRORIST ATTACKS UPON THE UNITED STATES, 9/11 AND TERRORIST TRAVEL: A STAFF REPORT OF THE NATIONAL COMMISSION ON TERRORIST ATTACKS UPON THE UNITED STATES (Hillsboro Press 2004) [hereinafter *9/11 and Terrorist Travel*].

2. NATIONAL COMMISSION ON TERRORIST ATTACKS UPON THE UNITED STATES, 9/11 COMMISSION REPORT: FINAL REPORT OF THE NATIONAL COMMISSION ON TERRORIST ATTACKS UPON THE UNITED STATES (W.W. Norton & Co. 2004), p. 384 [hereinafter *9/11 Commission Report*].

3. *Id.*

4. Unless otherwise noted, details of the structure and operation of the US-VISIT program have been collected from: (1) the Department of Homeland Security Web site (www.dhs.gov) as of September 2005; (2) Implementation of the United States Visitor and Immigrant Status Indicator Technology Program (US-VISIT), Biometric Requirements, 69 Fed. Reg. 468 (Jan. 5, 2004); and (3) US-VISIT, 69 Fed. Reg. 53318 (Aug. 31, 2004).

5. The additional exit points are: Atlanta, Georgia (William B. Hartsfield International Airport), Dallas/Fort Worth, Texas (Dallas/Fort Worth International Airport), Denver, Colorado (Denver International Airport), Detroit, Michigan (Detroit Metropolitan Wayne County Airport), Ft. Lauderdale, Florida (Ft. Lauderdale/Hollywood International Airport), Newark, New Jersey (Newark International Airport), Philadelphia, Pennsylvania (Philadelphia International Airport), Phoenix, Arizona (Phoenix Sky Harbor International Airport), San Francisco, California (San Francisco International Airport), San Juan, Puerto

Rico (Luis Muñoz Marin International Airport), Seattle, Washington (Seattle/Tacoma International Airport), and Los Angeles, California (San Pedro and Long Beach Seaports).

6. Established in 1986, the Visa Waiver Program allows nationals of the following countries to enter the United States for up to 90 days without a visa: Andorra, Iceland, Norway, Australia, Ireland, Portugal, Austria, Italy, San Marino, Belgium, Japan, Singapore, Brunei, Liechtenstein, Slovenia, Denmark, Luxembourg, Spain, Finland, Monaco, Sweden, France, Netherlands, Switzerland, Germany, New Zealand, and the United Kingdom. Visitors entering the United States under this program must now have machine-readable passports to avoid requiring a visa.

7. Press Release, U.S. Department of Homeland Security, Fact Sheet: U.S.-Mexico Land Borders, *available at* http://www.dhs.gov/interweb/assetlibrary/US-VISIT_Mexico_Fact_Sheet-English.pdf.

8. *Id.*

9. *Id.*

10. *Id.*

11. Brian Donohue, *Four Years After 9/11, U.S. Still Can't Track Visitors With Expired Visas*, NEWHOUSE NEWS SERVICE, Sept. 10, 2005.

12. The primary statutory provisions authorizing the US-VISIT program are: (i) Section 2(a) of the Immigration and Naturalization Service Data Management Improvement Act of 2000 (DMIA), Pub. L. No. 106–215; (ii) Section 205 of the Visa Waiver Permanent Program Act of 2000 (VWPPA), Pub. L. No. 106–396; (iii) Section 414 of the Uniting and Strengthening America by Providing Appropriate Tools Required to Intercept and Obstruct Terrorism Act of 2001 (USA PATRIOT Act), Pub. L No. 107–56; and (iv) Section 302 of the Enhanced Border Security and Visa Entry Reform Act of 2002 (Border Security Act), Pub. L. No. 107–173.

13. Implementation of the United States Visitor and Immigrant Status Indicator Technology Program (US-VISIT); Biometric Requirements, 69 Fed. Reg. at 468.

14. Department of Homeland Security, US-VISIT Program Privacy Impact Assessment Update 4-5 (June 15, 2005); Allan Holmes, *Cheap, Fast or Secure—Pick Two*, CIO MAG. (Sept. 2005).

15. In July 2005, DHS Secretary Michael Chertoff announced that US-VISIT would begin to include a 10-fingerprint scan as part of the visa application process.

16. Immediately after the September 11 attacks, the Department of Justice created a national registry of foreign visitors called the NSEERS, which was a first step toward the creating of US-VISIT. NSEERS involved fingerprinting visitors, periodic re-registration and checking out upon exit from the country. The program initially applied to all non-immigrant aliens from certain specified countries (such as Iran, Iraq, Libya, Sudan and Syria) and certain other individuals. Although the re-registration requirements of NSEERS no longer apply, some of the provisions are still operable. For those who are subject to both NSEERS and US-VISIT, complying with the entry and exit registration requirements of NSEERS will satisfy the requirements of US-VISIT.

17. *9/11 and Terrorist Travel*, *supra* note 1, at 7.

18. Press Release, Department of Homeland Security, Fact Sheet: US-VISIT (June 1, 2004), *available at* http://www.dhs.gov/dhspublic/display?content=3689.

19. Implementation of US-VISIT, Biometric Requirements, 69 Fed. Reg. 468, 477 (Jan. 5, 2005).

20. *See, e.g., 9/11 Commission Report*, *supra* note 2, at ch. 13.3.

21. Implementation of US-VISIT, Biometric Requirements, 69 Fed. Reg. at 477.

22. Implementation of US-VISIT, 69 Fed. Reg. at 471.

23. *Id.*

24. *Id.*

25. For a more detailed description of the protective measures used, *see* Privacy Impact Assessment Update for the US-VISIT Program (July 1, 2005), *available at* http://www.dhs.gov/interweb/assetlibrary/privacy_pia_usvisitupd1.pdf.

26. Press Release, U.S. Department of Homeland Security, Department of Homeland Security Announces Appointment to Data Privacy and Integrity Advisory Committee (Feb. 23, 2005), *available at* http://www.dhs.gov/dhspublic/interapp/press_release/press_release_0625.xml.

27. *Id.*

28. Press Release, Office of the Press Secretary, US VISIT Begins Deployment of Biometric Entry Procedures to Additional Land Border Ports of Entry with Canada and Mexico (Sept. 26, 2005), *available at* http://64.233.161.104/search?q=cache:VPC2kuiozbwJ:homelandsecurity.osu.edu/focusareas/sensors.html+elaine+dezenski+immigration+violators&hl=en.

Chapter 4

Biometrics and National Identification Cards

Taryn Lam and Cynthia-Clare Martey*

I. Introduction

"Identity (ID) cards are in use, in one form or another, in numerous countries around the world. The type of card, its function, and its integrity vary enormously. Around a hundred countries have official, compulsory, national IDs that are used for a variety of purposes. Many developed countries, however, do not have such a card. Among these are the United States, Canada, New Zealand, Australia, Ireland, the Nordic countries, and Sweden. Those that do have such a card include Germany, France, Belgium, Greece, Luxembourg, Portugal and Spain."[1]

Participation in national identification card schemes may be mandatory or voluntary. Traditionally, such programs have been used primarily to control access to government services and to determine the right to pass through a security checkpoint, such as border control at ports of entry into a country. These cards usually contain identifying information such as name, date of birth, gender, and a government-issued identification number. In some countries additional information such as height, eye and hair color, or current address can be included as well. Despite the name, national identification cards are not necessarily cards, but rather come in a variety of formats—plastic cards, certificates, booklets, etc. For example, Afghanistan uses a 16-page booklet

---

* Taryn Lam and Cynthia-Clare Martey are associates at White & Case, LLP, Palo Alto, California.

called a *taskera* or a *taskira*, and Poland issues a booklet that is similar to the U.S. passport.

The tragedies that occurred in the United States on September 11, 2001, reenergized the ongoing debate over the usefulness of national identification cards as a way to improve national security and to guard against terrorism. Particularly, the use of biometrics in such schemes has been fiercely debated in various countries around the world. This article will describe how countries have used, are using, or are debating the use of such cards, will examine the integration of biometric technology into the cards, and will discuss the policy and legal reasons behind the adoption or rejection of biometrics on national identification cards.

II. Worldwide Use of National Identification Cards

Approximately 100 countries have national ID card systems.[2] Use of national ID cards is not confined to a particular type of country or government or to a particular area of the world. The countries that utilize national ID cards span almost every continent and include both developed and third-world countries. Examples include Spain, Portugal, Germany, France, Italy, Belgium, Greece, Luxembourg, Poland, Pakistan, China, Singapore, and Thailand. In contrast, the United States, Canada, Mexico, the United Kingdom, New Zealand, Ireland, Sweden, and Bangladesh do not currently utilize such cards. Table A contains a non-exhaustive list of countries that have implemented national ID systems.

The following subsections describe a few of the countries that use national ID cards.

A. Brazil

The Brazilian identification card is compulsory and must be carried at all times. The card contains names of the bearer and his or her parents and the bearer's national status, photograph, thumbprint, and a serial number.[3]

B. Chile

Chile's residents carry a small plastic card that contains the bearer's name, date and place of birth, photograph, signature and personal identification number.[4]

C. China

The Chinese national identification card displays the cardholder's photograph, name, sex, ethnic nationality, birth date, address, date of issue and years of validity, and an identification number.[5] The card is issued by the Public Security Bureau. In the early 1990s, the government began issuing identification cards that contained holograms; in June 2001 microchips were added to cards issued by some cities.[6]

D. Croatia

A Croatian resident carries a card that displays her photograph, date and place of birth, address, serial and registry number and signature.[7]

E. Germany

The German national ID card displays the cardholder's name, date and place of birth, nationality, address, height, eye color, signature and photograph.[8] The card also contains its date of issuance, date of expiration, the issuing authority, and card number.[9]

F. Israel

The Israeli government requires that all citizens over the age of 16 carry a national identification document.[10] The document contains the holder's personal information and photograph.[11] This document is used for official identification purposes but a passport is still required for international travel.[12]

G. Italy

Italian residents carry a large card that contains the cardholder's name, date and place of birth, and photograph.[13]

H. Pakistan

The Pakistan ID card contains the name of the cardholder and his or her father, along with the cardholder's address, date of birth, photograph, and signature.[14] The card also carries the issuing official's signature, the date of issuance, the card serial number, and the ID Card number.

III. Integration of Biometrics into National Identification Cards

Biometrics technology is a means of identifying a person or verifying the claimed identity of a person using unique physical or behavioral characteristics. Fingerprints, facial recognition, and iris scans are some of the biometrics commonly found in biometrics applications. Although biometrics may conjure up futuristic images as depicted in Hollywood movies, biometry or the application of statistical analysis to biological data is a well-established science. Modern computing has allowed biometrics to be increasingly employed in the form of pilot projects in immigration settings such as airports, land border crossings, and other service points. As an emerging technology, biometrics is beginning to be used in a range of public- and private-sector applications, such as hospitals and banks.

The integration of biometrics into national ID cards is a controversial topic that has been or is being debated in several countries. Proponents of the idea argue that by including an individual's fingerprint or retinal pattern in his ID card, the card will be more effective. The verification of a person based on his identification cards is likely to be much more accurate if the verification results from a technologically advanced fingerprint or retinal scan rather than a government worker's cursory comparison of the person's photograph or signature.

The increasing popularity of biometric technology may result in the utilization of the technology in more and more national ID cards. As discussed below, it is a real possibility that beginning in 2008, American citizens could be required to carry national identification cards that contain biometric technology. The European Justice and Home Affairs Counsel recently endorsed the inclusion of an electronic chip containing facial and fingerprint biometrics in the national ID cards of the European Union's member states.[15] The national ID cards proposed by the government of the United Kingdom also include biometric technology. In October 2005, Australia introduced biometric e-passports—one of the first countries in the world to do so. A few years ago, Malaysia and Hong Kong Special Administrative Region introduced biometrics to their ID card schemes. The following case study will examine Malaysia's MyKad smart ID card, Hong Kong's smart identity card, and the Australian biometric e-passport programs as examples of how biometric technology may be integrated into ID cards.

A. Countries Using Biometrics In Their National Identification Card Schemes

1. Malaysia: MyKad Smart ID Card

Malaysia is one of the many countries that have had a national ID card program in place for many decades. Since 1949, all Malaysians over the age of 12 were required to register for a card. In September 2001, Malaysia launched a multipurpose smart ID card—the MyKad.[16]

The MyKad, as the first government-backed smart-card initiative in the world, crystallizes the major advantages and concerns presented by ID card schemes that utilize biometric information. The intention of the then prime minister, Dr. Mahathir, was to consolidate the many cards that existed for various functions into a single card incorporating all relevant information.[17] The MyKad was originally intended to include national ID card information, drivers license information, passport information, and health care information.[18] Instead, as implemented, the MyKad includes all the information originally intended as well as electronic purse information, ATM access information, public transportation features, and public key infrastructure (PKI) information for digital signatures that allow secure online transactions.[19] Thus, with a single swipe of his card, the Malaysian resident is able to have his identity verified, receive authorization to drive a vehicle, travel out of the country, have a record of his allergies and blood type, pay for small purchases, withdraw cash from his bank account, pay for his bus ticket, or transmit information online using his digital signature.

The wealth of information available on the MyKad, and the fact that the information is compressed into one single location, the card, make the erosion of personal privacy as a result of the use of the MyKad a serious concern. The MyKad displays the standard personal information found on the old identity card, driver's license, and passport.[20] In addition, the Mykad also displays the cardholder's gender and religion (if Muslim). The chip contains information about, *inter alia*, the cardholder's race and religion (regardless of the religious affiliation), thumbprint, voter registration, criminal record, driving demerits, and medical history.[21] Thus, it is clear that the MyKad contains significant amounts of personal, sensitive, and biometric information. It is also clear that the uses and functions of the MyKad are myriad and continue to expand.

Officially, the MyKad is only "optional" for Malaysians. However, because Malaysia's National Registration Department ceased issuance of the old identity cards in July 2001, anyone who has lost or damaged the old card or who must apply for an identity card must apply for a MyKad.[22] Therefore, eventually, every Malaysian resident will hold a MyKad.

2. Hong Kong: A Case Study

In October 2000, the Hong Kong Special Administrative Region announced its intention to replace the national identification card with a smart card that uses biometric technology to verify identity.[23] At the time, a compulsory national identification card regime already existed. A law called the Registration of Persons Ordinance (RPO) required that every person over the age of 15 register with the government and carry at all times th Hong Kong Identity Card (HKID card).[24] Although the RPO was enacted in 1949, at the time of its enactment, it only required registration.[25] During 1979 and 1980, in an effort to combat illegal immigration, the government changed the law to compel all persons to carry the identification card at all times.[26] The police may stop a person in the street at any time to check his HKID card; no probable cause of any kind is needed.[27]

The Immigration Department is in charge of issuing the HKID cards. Under the government's plan, the replacement of the old HKID cards with the new smart cards was to occur in two phases, with the first phase beginning in August 2003.[28] All residents are expected to have received the new cards by the end of 2007.[29] The new card will display the cardholder's name, birthdate, and immigration status. The card's embedded silicon chip will contain the cardholder's digitized photograph and an algorithm of his or her thumbprints. To verify identity, the cardholder places his thumb on an optical scanner, which converts the thumbprint into byte code. The byte code is compared against the fingerprint algorithm stored in the smart card's chip, and identity is established if there is a match.

The new smart card is to be used to verify identity and immigration status. Like Malaysia's MyKad, the Hong Kong smart ID card can be used for applications other than those for immigration purposes—such as driver's license, library card, and security certificates for digital signatures.[30] The capacity for an electronic purse application is also under development. However, unlike Malaysia's MyKad, with the Hong Kong smart ID, those non-immigration applications are optional and the information required for the optional applications are not included on the card unless the holder requests those applications.[31] The Hong Kong government has plans to expand the uses of the card to include automatic voter registration or automated passenger clearance at airports.[32]

The Hong Kong government presented a variety of reasons to justify using smart-card technology for its new ID card system. One hope is that the use of the smart cards will result in greater convenience to residents and will promote the use of technology.[33] The cards are also thought to be more resistant to forgery and identity theft.[34] The government also claimed that the smart cards would aid in border control and are necessary to replace Hong Kong's existing, unsustainable identification card system.[35]

3. *Australia: ePassport*

While this is a passport program and not a national identification card program, Australia's ePassport program represents a major shift in passport technology by its introduction of computer chips and biometrics.[36] Moreover, in a country where in 1986 a proposal for national ID cards was soundly defeated by popular revolt, the current acceptance of the newly proposed national identification card scheme may be due in part to the current anti-terrorism efforts by various world governments, and possibly in part to the ePassport program.[37]

The ePassport generates facial biometric information by digitizing the photograph provided by the passport applicant. The digitized photo is stored on the government's passport database and on the computer chip embedded in the middle pages of the ePassport. The chip also stores the name, sex, date of birth, place of birth, nationality, passport number, and the passport expiry date.[38] Unlike the MyKad and the Hong Kong smart ID card, the Australian ePassport currently has no function other than as a passport.

B. Countries Debating National Identification Card Schemes Including Biometrics

The remainder of this section will describe the recent developments and debates in the United States, the United Kingdom, and Canada regarding the implementation of a national identification card scheme.

1. *The United States*

Currently, the United States does not have a national identification card system, although there is room to argue that the driver's license has become a de facto national ID card. However, because driver's licenses are issued by the states, rather than by the federal government, the licenses vary in their contents and requirements.

This will change in the very near future. In May 2005, Congress and President Bush approved an $82 billion military spending bill. One of the most controversial portions of the bill is called the Real ID Act.[39] The Real ID Act essentially mandates the implementation of a federal identity card by May 11, 2008. The new law requires that state driver's licenses and other state identification cards comply with a uniform, federal set of standards. The Department of Homeland Security will determine these standards, as well as the other details of the new law's requirements.

Under the Real ID Act, U.S. citizens will be required to display their identification card in order to travel by air, to open a bank account, to enter a federal building, or to gain access to almost any government service.[40] Each card will contain, at least, the cardholder's name, date of birth, gender, address, photograph, and identification number.[41] Because the states' databases would be linked to one another, a person's record would follow him or her even after moves from one state to another.[42]

The Act requires the cards to be "machine-readable" but does not specify the technology that must be used. The Department of Homeland Security may choose to add a magnetic strip, a bar code, a radio frequency identification chip, DNA information, or biometric information such as a fingerprint or retinal pattern.

The main impetus behind the Real ID Act was the September 11 attacks. The fact that the terrorists possessed legitimate driver's licenses exposed the vulnerabilities of the existing identification regime. Proponents of the Act argue that it will prevent terrorists and other criminals from moving freely throughout the country. Opponents decry the Act as further government intrusion into personal privacy.

The U.S. Department of Homeland Security is currently conducting a second test of e-passports containing a microchip encoded with *inter alia* biometric information at select airports in the United States, Asia, and Australia for the purpose of gathering information to help countries develop electronic passports.[43] The results of such tests bring the United States closer to the use of biometrics at least on passports, and possibly as part of a national identification card scheme.

2. The United Kingdom

Recently, the UK was embroiled in rich debate regarding the planned introduction of passports and/or identity cards that use biometrics and that are linked to a national identity database.[44] This national identity database has been the subject of much criticism. Such a national database would allow identification of persons whose information is part of the database even if they do not carry a biometric-enabled card, provided they are willing to submit to a scan.

Like the United States, the United Kingdom does not currently have a national identification card system in place. However, in 2003 the British government announced its intent to pass a bill that would create a British national identity card. The bill, titled the Identity Cards Bill,[45] further provides for a two-stage implementation of the cards. Initially, the cards would be voluntary but would become compulsory for all residents by 2013.

The Identity Cards Bill would link the card to a national database called the National Identity Register. Together, the card and Register would record the biometric information belonging to UK residents, such as fingerprints and facial and retinal scans. Despite the title of the bill, the most important and controversial component of the bill is the Register, not the identification card itself. Because biometric information is stored in the Register, identification and verification of identity can be accomplished with a biometric scan, without checking the identification card itself.

An earlier version of the Identity Cards Bill was introduced in the House of Commons on November 29, 2004.[46] Although that version was approved by the House of Commons, there was insufficient time for the bill to be debated and approved in the House of Lords prior to the dissolution of the Parliament on April 11, 2005.[47] After its electoral victory in 2005, the Labour Government reintroduced a new version of the bill, which was substantively the same as the earlier version, to the House of Commons on May 25, 2005.[48] Both the Liberal Democrat and Conservative parties opposed the bill.[49] Despite the opposition, the bill passed in the House of Commons in its second and third readings.[50] However, the House of Lords defeated the bill by instead supporting a fully voluntary ID card system wherein an individual's entry of his or her personal information onto the Register is voluntary and the transition from a voluntary ID card to a compulsory card would require Parliamentary approval rather than occurring automatically.[51] In 2006, the British government finally passed the Identity Cards Act 2006. Under the act, in 2008, all British citizens who renew their passports will be issued an identity card and will be added to the Register. However, until 2010, citizens may decline issuance of the identity cards.

According to the UK government, implementation of the Identity Cards Bill will help combat identity theft, detect illegal immigrants and illegal workers, prevent misuse of public services, and guard against terrorism and organized crimes.[52] Despite the government's assurances, the idea of one centralized database containing the personal information and biometric data of every British citizen has raised protests regarding data protection and privacy concerns.[53]

3. Canada

In the aftermath of the September 11, 2001, attacks in the United States, the Canadian federal government has considered implementing a national identification card scheme integrating biometric technology, in part to strengthen national security and to help corporations fight identity theft. The debate is ongoing and centers on many of the same issues discussed in the section below.

IV. The Policy and Legal Debate Raised by the Use of Biometrics on National ID Cards

The debate regarding the use of biometrics as part of the national identification card scheme often centers on the tension between the technological advantages gained from the use of biometrics and the various policy and legal issues raised by the use of the technology. This chapter does not address the technological issues—which are not without some controversy and debate. Instead, the remainder of this chapter highlights some of the policy and legal issues that have been and continue to be debated around the globe.

The main benefit of a national identification card system is that the identity can be established through a single, official document that serves as the repository for all of an individual's personal information. As a result, the verification of the individual's identity is easier, more efficient, and more accurate. The addition of biometric technology to the cards can further increase their accuracy. Once established, however, it is not unusual for government to use national ID cards for identification as well as other purposes. Often, it is these other purposes that raise the ire of those who oppose the use of biometrics.

One purpose behind the use of a national identification card system is greater national security and better immigration and border control. The cards allow a government to detect persons, such as terrorists and illegal immigrants, who are not entitled to be inside the country. The cards could also be used to limit access to government services, such as welfare or healthcare benefits, to only those who are entitled to such benefits. In addition, voting fraud may be reduced if voter registries are based on national ID cards. Many of these advantages are defined by the inherent advantages and disadvantages of biometric technologies that are beyond the scope of this chapter.

A. Privacy

Privacy is the primary issue raised in all the debates surrounding the implementation of national identification cards that utilize biometric technology. Looking at the amount and type of personal, sensitive, and biometric information stored on cards such as the MyKad, the Hong Kong smart ID card, and the Australian ePassport, it appears to be a justified concern. The concern is highlighted by the existence of the national database of identification and other information, such as that proposed by the UK, which must exist to support any biometric-enabled ID card scheme.

The privacy issues raised by these matters have been researched and described often. The concerns arise for several reasons, including the following: (i) identity checks increase the contacts individuals have with government agents; (ii) national ID cards enhance government access to information about a person's activities, associations, etc. A national identity card system therefore affects both informational privacy (the freedom to limit access to one's personal information) and physical privacy (the freedom from physical contact with others).[54] Many countries have laws protecting the privacy of their citizens. The concern is that increased contacts and increased access would have a chilling effect on the legal exercise of such personal freedoms.[55]

For example, the Fourth Amendment to the United States Constitution confers the right to be free from unreasonable searches and seizures. A biometrics-based national ID card system would raise questions regarding, for example, when a law enforcement official may request a biometric scan to verify identification, or if every person subject to an investigative or traffic stop can be compelled to submit to a biometric scan. Such a system would also raise issues about the extent of an individual's right to refuse to submit to a biometric scan under the Fifth Amendment's protection against self-incrimination.

B. Function Creep

Another concern arises from the potential use of personal information in the system for purposes not originally intended. As previously discussed, the primary functions of national identification schemes are to provide a means to qualify people for access to government services and to pass security checkpoints. However, as demonstrated by the MyKad and the Hong Kong smart ID cards, the technology lends itself to many applications, and therefore it is quite likely that such cards will wind up being used not only for the government-sanctioned purposes, but also for ever-increasing public and private applications.

This slippery slope applies particularly to the increasing use by the government of such powerful tools for pettier functions unrelated to national security or immigration.[56]

Such expanding use is not necessarily wrong or problematic in itself. However, with increasing applications, the number of uses that are not regulated by the government increases. Absent the regulation of government, any constitutional or other regulatory protections against misuse, abuse, or even discrimination may go unchecked.

There are increased opportunities for discrimination on the basis of biometric characteristics or, in voluntary national ID card schemes, for discrimination against those who choose not to participate in all or part of the scheme. For example, the owner of a copy shop may offer discount pricing to those who use the electronic wallet/purse feature of the MyKad, thus resulting in price discrimination against those who do not have a smart card (either by choice or because they are illegal immigrants and do not qualify for a card) for something as simple as buying a cup of coffee.

C. Discrimination

Many national identification cards display the cardholder's race, ethnicity, or religion. The danger in this is that such information can be used to discriminate against the cardholder. For example, a government official may be more likely to detain, interrogate, or abuse an individual when the official learns that the individual is a member of a certain ethnic group. Taken to the extreme, group classifications on the cards may facilitate horrors such as genocide.[57]

V. Conclusion

The above and other matters will need to be discussed in legislative, academic, and public circles before there can be any resolution of this pressing debate. However, given the present state of worldwide terrorism and the desire of various governments to shore up their borders and to improve interior defenses, it is likely that many more countries will be following in the footsteps of Malaysia and Hong Kong in the months and years ahead. Indeed, even countries that do not currently have national identification cards will likely adopt schemes that use biometric information.

Table A: Non-exhaustive List of Countries That Utilize National ID Cards[58]

Name of Country	Name of ID Card	Year Established
Afghanistan	Taskera/taskira	1973
Belgium	Carte d'identite / identiteitskaart	N/A
Bhutan	Citizenship identity card (CIC)	1958, 1985
Bosnia/Herzegovina	Licna Karta	N/A
Brunei	N/A	N/A
Burma	N/A	1990 (new format)
Burundi	Carte d'identite	1930s; discontinued in 1962
Cambodia	Special identity certificates	1993
China	Jumin shenfenzheng	1985
Croatia	Osobna iskaznica; osobna karta	1991
Chile	N/A	N/A
Dominican Republic	Cedula personal de identidad	N/A
Egypt	N/A	N/A
Ethiopia	N/A	N/A
France	Carnet anthropometrique	1912, revised in 1970
Georgia	N/A	1996
Germany	Personalausweis	N/A
Greece	Deltio Taytotitas	N/A
Indonesia	Kartu tanda penduduk	N/A
Iraq	N/A	N/A
Israel	Teudat Zehut	1949, 1958
Italy	N/A	N/A
Japan	N/A (alien registration cards)	N/A
Jordan	N/A	N/A
Kenya	Kitambulisho; kipande	N/A
Korea	National registration card	N/A

Name of Country	Name of ID Card	Year Established
Laos	N/A	N/A
Lebanon	Tathkarat al-Hawiya or idhraj kayd qayd	1970s
Luxembourg	N/A	N/A
Macedonia	Licna karta	1995
Malaysia	National Registration Identity Card	1949
Pakistan	National Identity Card	1973
Poland	Dowod osobisty	N/A
Portugal	N/A	N/A
Russia	Identity documents and propiska	1932
Rwanda	Carte d'identite (resident permits)	1933; 1962; discontinued in 1996
Saudi Arabia	National ID cards and iqamas (resident ID cards)	N/A
Serbia/Montenegro	Licna karta	N/A
Singapore	National Registration Identity Card	N/A
Slovenia	Osebna izkaznica	N/A
South Africa	Reference books	1891, 1952, 1966, reformed in 1986
Spain	N/A	N/A
Sri Lanka	National Identity Card	1972
Syria	N/A	N/A
Thailand	N/A	N/A
Turkey	Nufus Cuzdani; kimlik karty	N/A
Vietnam	N/A	1975

Notes

1. Simon Davies, *Identity Cards: Frequently Asked Questions, Privacy International,* Aug. 24, 1996, http://www.privacy.org/pi/activities/idcard/idcard_faq.html

2. *Id.*

3. *Id.*

4. *Id.*

5. *See Prevent Genocide International,* http://www.preventgenocide.org/prevent/removing-facilitating-factors/IDcards/survey/ (last visited Feb. 3, 2006).

6. *Id.*

7. *Id.*

8. Davies, *supra* note 1.

9. *Id.*

10. Prevent Genocide International, *supra* note 5.

11. *Id.*

12. *Id.*

13. *Id.*

14. Davies, *supra* note 1.

15. John Lettice, *EU Ministers Approve Biometric ID, Fingerprint Data Sharing,* THE REGISTER, Dec. 1, 2005, http://www.theregister.co.uk/2005/12/01/jahc_biometric_id_standards/.

16. Mathews Thomas, *Is Malaysia's MyKad the 'One Card to Rule Them All'?" The Urgent Need to Develop a Proper Legal Framework for the Protection of Personal Information in Malaysia,* 28 MELB. U. L. REV. 474, 475 (August 2004).

17. *Id.* at 476.

18. *Id.* at 481.

19. *Id.*

20. *Id.* at 481.

21. *Id.*

22. *Id.* at 483.

23. Rina C. Y. Chung, *Hong Kong's "Smart" Identity Card: Data Privacy Issues and Implications for a Post-September 11th America,* 4 ASIAN-PAC. L. & POL'Y J. 518, 521 (June 2003).

24. *See id.* at 527-28.

25. *See id.* at 528.

26. *See id.*

27. *Id.*

28. *Id.* at 537.

29. *Id.*

30. *Id.* at 521-522.

31. http://www.smartid.gov.hk/en/app/index.html (last visited Feb. 7, 2006).

32. *Supra* note 23 at 538.

33. *Id. at* 533.

34. *Id.*

35. *Id.* at 535.

36. http://www.passports.gov.au/Web/ePassport.aspx (last visited Feb. 7, 2006).

37. http://www.theaustralian.news.com.au/common/story_page/0,5744, 18002768%25255E2702,00.html (last visited Feb. 7, 2006).

38. *Supra* note 37.

39. The full text of the Real ID Act can be accessed at: http://thomas.loc.gov/cgi-bin/query/z?c109:H.R.418:

40. Declan McCullagh, *FAQ: How Real ID Will Affect You,* CNET, http://news.com.com/FAQ+How+Real+ID+will+affect+you/2100-1028_3-5697111.html (last visited Feb. 6, 2006).

41. Wikipedia, http://en.wikipedia.org/wiki/Real_ID_Act (last visited Feb. 6, 2006).

42. A state must agree to link its databases in order to receive federal funds for the program. McCullagh, *supra* note 40.

43. http://news.zdnet.com/2100-1009_22-6034384.html (last visited Feb. 6, 2006).

44. U.K. Identity Cards Bill, HL Bill 28, Oct. 19, 2005.

45. The full text of the most recent version of the bill is available at the United Kingdom Parliament's Web site: http://www.publications.parliament.uk/pa/ld200506/ldbills/071/2006071.htm (last visited Feb. 6, 2006).

46. Wikipedia, http://en.wikipedia.org/wiki/British_national_identity_card (last visited Feb. 6, 2006).

47. *See id.*

48. *See id.*

49. *See id.*

50. *See id.*

51. *ID Cards Scheme in Lords Defeat*, BBC NEWS, http://news.bbc.co.uk/1/hi/uk_politics/4640900.stm.

52. Home Office (The United Kingdom), Identity Cards Briefing, May 2005, *available at* http://www.identitycards.gov.uk/library/Id_Cards_Briefing.pdf.

53. *Deal Paves Way for ID Cards*, BBC News, http://news.bbc.co.uk/1/hi/uk_politics/4856074.stm (last visited Nov. 16, 2006).

54. John D. Woodward, *Biometric Scanning, Law & Policy: Identifying the Concerns—Drafting the Biometric Blueprint*, 59 U. PITT. L. REV. S10 (Fall 1997).

55. K. A. Taipale, *Technology, Security and Privacy: The Fear of Frankenstein, the Mythology of Privacy and the Lessons of King Ludd*, 7 YALE J.L. & TECH. 123 (2005).

56. *Id.*

57. *See* Prevent Genocide International, http://www.preventgenocide.org/prevent/removing-facilitating-factors/IDcards/survey/ (last visited Feb. 3, 2006).

58. *Id.*

Chapter 5

Do Biometric Identifiers Make Us Safer?

*Kyle D. Chen**

I. Introduction

Many people believe that biometric identifiers may provide tools to improve security because they are closely tied to individuals and are not easily lost, forgotten, falsified, or stolen. One key advantage of biometric identifiers is that they do not depend on an object (such as a card or a badge) or a piece of secret information (such as a password or a personal identification number). Instead, they rely on the unique biological information of people to identify them or to verify their identities.

A biometric system generally requires an "enrollment" process to create the basis for the comparisons required for identification or verification of identities. During the enrollment process, the system captures individuals' sample biometric identifiers and converts them into biometric templates, which are then stored in a database for later comparisons. When in operation, the system compares the presented biometric identifiers against the templates stored in the database to determine if they match.

In the identification mode, the system compares the presented biometric identifier against all templates stored in the database and answers the question "Who is this person?" This mode requires the system to conduct what is known as "one-to-many" comparisons. There are two types of

* Associate, White & Case LLP, Palo Alto, California.

identification systems, positive and negative. A positive identification system attempts to confirm that the presented biometric identifier matches one of the stored templates. An access control system for a building, for example, would be such a positive system. A negative identification system, by contrast, attempts to confirm that the presented biometric identifier matches none of the stored templates. An example of such negative system would be a terrorist watch list at the U.S. border.

In the verification mode, the system compares the presented biometric identifier against the template of a particular claimed identity to answer the question "Is the person who he or she claims to be?" This mode requires the system to conduct what is known as the "one-to-one" comparison. A system that checks the biometric identifier of the person with an identification card against the template of the person identified in the identification card would be an example of a biometric system in the verification mode.

Whether in the identification or the verification mode, a biometric comparison does not generate "perfect" matches. Instead, the comparison generates a score of how "close" the presented biometric identifier matches the stored template. If the score equals or exceeds a certain threshold, then the system will find a "match." If the score is below such threshold, then the system will find a "nonmatch."

The effectiveness of a biometric system heavily depends on its accuracy, for which two critical criteria are the "false-match rate" (FMR) and the "false-nonmatch rate" (FNMR). A false match occurs when the system recognizes a person presenting the biometric identifier as matching a stored template when he or she in fact did not produce the template matched by the system. The FMR is the probability of such false-match occurrences. A false nonmatch occurs when the system fails to recognize a person as matching a stored template when he or she in fact did produce the particular template. The FNMR is the probability of such false nonmatch occurrences.

The FMR and FNMR of a biometric system directly affect its capability to identify or verify individuals. In the positive identification or the verification mode, a false match results in one being incorrectly granted access when his or her access should have been denied, while a false nonmatch results in one being wrongly denied access when his or her access should have been granted. In the negative identification mode, a false match results in one being erroneously identified as a targeted or watched individual, while a false nonmatch results in a targeted or watched individual escaping monitoring. Depending on the types of biometric identifiers and the thresholds employed, different biometric systems will have different FMRs and FNMRs. Both the FMR and the FNMR of a well-performing biometric system should be low.

In addition to the FMR and the FNMR, a third important criterion for accuracy in measuring a biometric system's effectiveness is the "failure-to-enroll rate" (FTER). The FTER of a biometric system is the percentage of people failing to enroll. A person may fail to enroll because he or she cannot provide adequate biometric samples. For example, the fingerprints of construction workers may often be too worn to be captured. A biometric system's performance also heavily depends on its FTER, because a high FTER means that the system simply cannot identify or verify the identities of a large percentage of people.

II. Effectiveness of Various Biometric Systems in Practice

Biometric identifiers appear to provide useful tools for identification and verification, but in order to determine whether they actually improve safety, we need to examine the effectiveness of biometric technologies in practice. The foremost consideration is certainly accuracy (i.e., the system's FMR, FNMR, and FTER). Many other issues, such as costs, intrusiveness, and user acceptance, are also important considerations. In the following sections, we will investigate the effectiveness of seven biometric technologies, including facial recognition, fingerprint recognition, hand geometry, iris recognition, retina recognition, voice recognition, and signature recognition.

A. Facial Recognition

Facial recognition technology analyzes facial features that are not easily altered, including the upper outlines of the eye sockets, the areas around the cheekbones, and the sides of the mouth. The system may compare a stored template against the live scan of a person's face or a static image, such as a person's photograph on a passport. The technology operates in both the identification and the verification modes. In addition, because of the ease in capturing the image of a person's face, facial recognition technology has very low FTER and is the only biometric technology suitable for surveillance purposes.

The overall effectiveness of facial recognition technology has been less than ideal. The accuracy of facial recognition heavily depends on environmental factors, especially lighting conditions. Camera variations, facial positions, expressions, aging, and alterable facial features (e.g., hair styles, beards, etc.) further degrade its accuracy. Based on data from pilot programs of facial recognition systems deployed at airports, the FMRs ranged between 0.3 percent and 5 percent, and the FNMRs ranged between 5 percent and 45 percent.[1] In a State Department Bureau of Consular Affairs test, the FNMRs have been reported to be as high as 70 percent.[2]

The high FNMRs of facial recognition systems may severely limit their effectiveness for certain applications. In the verification mode or the positive identification mode, for example, the systems' high FNMRs may translate into denial of access to a large percentage of people who should have been granted access. In the negative identification mode, high FNMRs may result in many people being granted access when their access should have been denied. The consequence of high FNMRs in the negative identification mode may be particularly dire in the context of, for instance, a terrorist watch list at the border, because dangerous individuals who should have been denied entry to the country would be able to enter, posing great risk to the national security.

Facial recognition systems can be expensive. For example, an access control server using facial recognition technology for a facility with up to 30,000 persons would cost about $15,000.[3] Depending on the number of installations, the software license would cost between $650 and $4,500.[4] Larger databases and more attempted matches would further increase the cost. A live-scan facial recognition system may also include closed-circuit television (CCTV), which can cost from $10,000 to $200,000, depending on the size of the entrance and the degree of monitoring.[5] Each additional CCTV camera costs between $125 and $500. A camera with advanced features may reach $2,300.[6]

User acceptance of facial recognition is reasonable because, typically, facial recognition is thought to be less intrusive than other biometric systems. Some worry, however, that facial recognition technology may track people without their consent. Despite the high cost and less than ideal accuracy, facial recognition remains attractive because of its minimally intrusive nature and its ease of use in operation, especially when utilized with facial images in a wide variety of identification documents.

B. Fingerprint Recognition

Fingerprint recognition is probably the most widely used and well-known biometric technology. Commercial applications of automatic fingerprint recognition have been available since the 1970s. Until recently, however, primary fingerprint recognition applications have been used for only law enforcement.

Fingerprint recognition technology takes advantage of the features extracted from impressions produced by distinct ridges on the fingertips. Two types of fingerprints are commonly employed in the technology: flat and rolled. A flat print utilizes only the impression of the central area between the fingertip and the first knuckle, while a rolled print covers the ridges on both sides as well. During enrollment, the system scans and enhances the fingerprints and converts them into templates, which are then saved into the database for later comparisons. Optical, silicon, or ultrasound scanners are available technologies for capturing fingerprint impressions.

The effectiveness of fingerprint recognition technologies is dependent on the type of application and the scanner employed. Generally, the FTER is about 2 percent to 5 percent.[7] The FMR may be set for different levels of security, with higher FMRs at lower levels of security (in the positive identification or verification modes). For instance, the FMR is 1.5×10^{-12} and FNMR 1.5 percent to 2 percent for the Integrated Automated Fingerprint Identification System (IAFIS) at the Federal Bureau of Investigation (FBI).[8] By contrast, in a test conducted by the Federal Aviation Administration (FAA) at airports from September 2000 to February 2001, the FNMRs were between 6 percent and 17 percent for closely controlled test subjects and between 18 percent and 36 percent for actual airport employees accessing doors in a less-controlled environment.[9] The FMRs were between

0 percent at the highest security level and about 8 percent at the lowest security level.[10]

Fingerprint recognition is relatively less expensive compared to other biometric technologies. Fingerprint readers designed for physical access control cost approximately $1,000 to $3,000 per unit.[11] The software license fee is about $4 per enrollment.[12] The maintenance cost is about 15 percent to 18 percent of the purchase price for smaller fingerprint scanners.[13] For larger live-scan 10-print fingerprint readers, the unit cost is about $25,000, for which the maintenance cost is about 14 percent of the purchase price.[14]

User acceptance of fingerprint recognition technology has been reasonable, although its traditional relationship to forensic fingerprinting in criminal investigations does result in some discomfort and resistance. Some people also worry that fingerprints captured for one purpose may be used to track them for other purposes. One reasonably successful application of fingerprint technology has been in veterans' hospitals.[15] The fingerprint recognition system has assisted in solving the challenge of access control in hospitals.[16] People have seemed comfortable with the system, and there have generally been very few complaints.[17]

C. Hand Geometry

Hand geometry technology distinguishes individuals by analyzing the widths, heights, and lengths of the fingers, distances between joints, and the shapes of the knuckles. The system captures two orthogonal two-dimensional images of the back and sides of the hand using an optical camera and light-emitting diodes with mirrors and reflectors. The system then conducts 96 measurements of these captured images and produces a template to store into the database.

Generally, hand geometry alone is not suitable for identification because the shapes and sizes of people's hands are reasonably diverse but not highly distinctive. The guiding pins on most hand readers can assist in properly positioning the hands to improve accuracy. The technology is useful in verifying the identity after, for example, the individual presenting the hand geometry has been identified by an identification card or by other biometric identifiers.

The FMR, FNMR, and FTER have been reasonably low for the hand geometry technology. In a test conducted by the FAA from March to July 2001 at airports, the FMR was between 0 percent at the high-security level and about 2 percent at the low-security level, while the FNMR was between 5 percent at the high-security level and less than 1 percent at the low-security level.[18] In addition, everyone with at least one hand can enroll in a hand geometry system; hence, the FTER is virtually zero.[19]

Other effectiveness considerations such as cost and user acceptance for the hand geometry technology have also been reasonable. The unit cost of a hand geometry reader usually ranges from $2,000 to $4,000.[20] The personnel cost is minimal, because most hand geometry readers can operate unattended. Generally, most people consider hand geometry systems to be nonintrusive, nonthreatening, and noninvasive. In addition, the hand geometry readers are not less hygienic than, say, common doorknobs.

D. Iris Recognition

Iris recognition technology takes advantage of the rich biometric data provided by the distinctly colored ring around the pupil of the eye. Among the 266 distinctive characteristics in the iris (including the trabecular meshwork with striations, rings, furrows, a corona, and freckles), 173 are typically employed to generate the templates. Barring injury, these characteristics of the iris reportedly remain stable throughout an individual's lifetime following their formation during the eighth month of gestation.

Iris recognition technology exhibits good accuracy characteristics. In a test conducted at the U.S. Army Research Laboratory, the FMR was less than 1 percent, and the FNMR was about 6 percent.[21] Another recent test conducted by the National Physical Laboratory (NPL) demonstrated an FMR of 0 percent and an FNMR of 0.2 percent.[22] In addition, NPL found the FTER of iris recognition technology to be 0.5 percent.[23] Furthermore, the actual experience of the United Arab Emirates (UAE) in employing iris recognition systems seems to indicate that this technology offers a useful means in preventing expelled foreigners from re-entering the country.[24] The statistical analysis has suggested an FMR of 1 in 80 billion, meaning practically no individuals *not expelled* have been erroneously identified.[25] However, little information is available to estimate the FNMR in such negative identification mode because tracking individuals who have escaped the monitoring of the UAE iris recognition system is, by definition, quite difficult.

Despite its superior accuracy characteristics, other considerations such as costs, optical factors, diseases, and user acceptance may somewhat affect the effectiveness of iris recognition technology. The unit cost for an iris recognition access point costs approximately $2,000, but the overall cost of a comprehensive system would be much higher.[26] Strong glasses and colored or bifocal contact lenses may interfere with the system performance, as may poor eyesight, glaucoma, or cataracts. In addition, some people find aligning their eyes steadily toward cameras very difficult, while others resist the scanning of their eyes.

E. Retina Recognition

Retina recognition technology takes advantages of the different patterns of blood vessels on thin nerves located at the back of the eyeball to identify individuals. Each eye exhibits highly distinctive retina patterns, even for identical twins. Generally, the retina patterns remain stable over lifetimes, but diseases such as glaucoma, diabetes, high blood pressure, and autoimmune deficiency syndrome may alter the patterns.

Retina recognition is considered to be an intrusive technology. The main reason is that the retina is small, internal, and hard to measure. Hence, capturing its image requires a great degree of effort and cooperation of users. As a result, although retina recognition is one of the most accurate and reliable biometric technologies, it is also one of the least deployed in practice. Today, retina recognition technologies are adopted only for extremely stringent access control in highly secure government and military environments, such as those at nuclear weapon and research sites.

F. Voice Recognition

Voice recognition technology utilizes differences in voices resulting from a combination of physiological differences in the shape of vocal tracts and learned speaking habits. The system enrolls an individual by having him or her speak some predetermined information, known as the passphrase, into a microphone or telephone several times. (Not all voice recognition systems require predefined passphrases, however.) The system then creates a voice template to store into the database by analyzing the individual's vocal characteristics, such as pitch, cadence, and tone. In practice, voice recognition may operate in both identification and verification modes.

Although its system is easy to use, minimally intrusive, and fairly inexpensive, voice recognition is not a very accurate biometric technology. Its reliability further degrades in noisy environments, such as those at points of entry. Generally, the technology is employed in situations in which a person's voice is the only available biometric identifier, such as those at telephone and call centers.[27]

G. Signature Recognition

Signature recognition technology measures characteristics such as the rhythm, acceleration, and pressure flow of handwritten signatures. Specifically, the system analyzes an individual's signature dynamics, including the absolute and relative speeds, stroke order, stroke count, and pressure. Based upon such analyses, the system then produces signature templates and stores them into the database for later comparisons. Generally, the technology operates in the verification mode rather than the identification mode.

III. Do Various Biometric Systems Make Us Safer?

Based upon the information presented above, biometric identifiers appear to provide a viable means to identify and verify people's identities in many contexts. Many currently available biometric technologies exhibit low FMRs, FNMRs, and FTERs in field tests with reasonable total costs, ease of use, and user acceptance. Potentially, these technologies may offer higher security beyond current means for critical infrastructures such as airports, governmental buildings, and the like.

However, as newly introduced security measures, if biometric identifiers operate as merely "additional" means for identification and verification, then there would appear to be little doubt that the technologies can only improve security. However safe we were before the introduction of biometric identifiers, the addition of these new security measures seemed to only improve the existing protection. All other things being equal, biometric identifiers would probably make us safer as merely an additional means for identification and verification.

The issue becomes more complicated if the introduction of biometric identifiers results in reduction of existing nonbiometric security measures. Biometric identifiers may replace certain nonbiometric security means thought to be "duplicative" of the biometric means for efficiency reasons. (For example, a security system may opt to replace identification documents altogether with the fingerprint technology operating in the identification mode.) As a result, depending on the actual performance of the biometric identifiers, the security level of the system may in fact increase or decrease. If the newly adopted biometric identifiers turn out to perform worse than the replaced measures, the system security may actually worsen. In addition, even assuming that the biometric identifiers actually work better, our overall safety may not improve if overreliance on biometric technologies results in reduced alertness and caution in people.

Another consideration is the possibility of deception in biometric technologies. It has been demonstrated that fooling a biometric system by presenting false biometric identifiers is actually possible. For example, a high-resolution photograph of an iris with a tiny hole cut out to permit the pupil of a live eye to shine through was shown to be able to deceive certain iris recognition systems.[28] To make the matter worse, once a biometric identifier such as the image of an iris is "stolen," it may never be reclaimed or recovered because of the uniqueness and unalterable nature of such identifier. Currently, there appears to be insufficient empirical data available for evaluating such deception issues, as pervasive deployment of biometric technologies for security is still quite limited.

IV. Conclusion

Biometric technologies seem to present useful tools to improve security. If they are employed in conjunction with existing security measures (for instance, biometric verification of one's claimed identify after his or her presentation of traditional identification documents), then there appears to be little doubt that such technologies would improve security. However, if biometric identifiers are to replace rather than to simply augment existing security measures, more empirical performance data of such biometric identifiers' field operations would be required to assess their reliability. In addition, we should still stay alert and exercise caution in maintaining our safety despite the advancement of biometric technologies. In particular, we should avoid overreliance upon biometric identifiers, especially because the likelihood of deception in biometric identifiers is yet to be fully understood.

Notes

1. *See* U.S. GENERAL ACCOUNTING OFFICE (GAO), TECHNOLOGY ASSESSMENT: USING BIO-METRICS FOR BORDER SECURITY 70 (GAO-03-174) (Nov. 2002).
2. *See id.*
3. *See id.*
4. *See id.*
5. *See id.*
6. *See id.*
7. *See* GAO, *supra* note 1, at 71.
8. *See* GAO, *supra* note 1, at 72.
9. *See id.*
10. *See id.*
11. *See id.*
12. *See id.*
13. *See id.*
14. *See id.*
15. *See* Paul Rosenzweig et al., Legal Memorandum (published by Heritage Foundation) No. 12, *Biometric Technologies: Security, Legal, and Policy Implications* 4 (June 21, 2004).
16. *See id.*
17. *See id.*
18. *See* GAO, *supra* note 1, at 74.
19. *See id.*
20. *See id.*
21. *See* GAO, *supra* note 1, at 73.
22. *See id.*
23. *See id.*
24. *See* Rosenzweig et al., *supra* note 15, at 3.
25. *See id.*
26. *See* GAO, *supra* note 1, at 73.
27. *See* Rosenzweig et al., *supra* note 15, at 5.
28. *See* GAO, *supra* note 1, at 73.

Chapter 6

Theft of Biometric Data: Implausible Deniability

Ian Johnson *

Proponents tout biometrics both as a surefire method for limiting access to restricted things and places and as a panacea for the increasing and increasingly bothersome incidence of identity theft.[1] Because it establishes identification through unique biological or physiological markers, they maintain, the technology all but eliminates the possibility of mistaking one person for another.[2]

The proponents have a point. If one wants to single Joe out from the rest of humankind, it makes perfect sense to focus on something in or on Joe that differs from the same thing in or on everyone else. And one cannot reasonably dispute that such differences exist, that certain physical traits simply do vary discernibly from one person to the next. The loops and whorls of his fingerprints, the vascular mosaic of his eyes, the precise geometry of his hand—each of these is certainly unique to Joe.

Of course, if it relied on mere uniqueness, biometrics technology would present no real advance on current identification technologies. Joe's ATM card and associated PIN differ from Jane's card and PIN just as certainly as his prints differ from hers. The advantage of biometrics, insist its champions, lies in the inimitability and immutability of the unique traits upon

* Ian Johnson is an associate with Orrick, Herrington & Sutcliffe LLP, San Francisco, California.

which it relies. Put another way, the superiority of biometrics rests on the assumption that certain biological and physiological traits never change and are exceedingly hard to fake.

This assumption gives rise to certain concerns—whether right or wrong. If it is right—if biometrics markers are, in fact, very difficult to replicate—fairly routine theft could conceivably take a turn toward the grisly. Suppose, for example, that fingerprint readers become a common security device on luxury cars. If prints cannot be forged or, at least, are perceived as being insusceptible to forgery, a thief bent on stealing a Mercedes might resort to cutting off the owner's finger. Lest this be viewed as fantastic, that very thing happened in Malaysia in March 2005. Shortly after leaving a bar in a Kuala Lumpur suburb, a Mr. Kumaran had lost his clothes, his Mercedes S-Class and the end of his index finger to thieves "frustrated" by their inability to bypass the car's fingerprint recognition system.[3]

As sad as Mr. Kumaran's plight is, it could have been worse. His Mercedes might have been equipped with a system designed to detect the veins in his left ear, left hand, or, worse, left iris. Still, while room for concern certainly exists, there is reason to believe that the world will be spared generations of one-handed or one-eyed luxury car drivers. Thieves will likely come to understand that amputated fingers and hands, severed ears and enucleated eyeballs have a short usable life. The cornea of an extracted eyeball, for example, quickly clouds, obstructing an electronic reader's access to the iris. Fingerprints too degenerate rapidly, becoming useless some 10 minutes after amputation.

The possibility that assumptions regarding the inimitability of biometrics markers are wrong—that is, the possibility that markers can be replicated with some regularity—gives greater pause. Suppose, for example, that some clever thief developed the wherewithal to lift Joe's fingerprints from Joe's coffee mug, to replicate those prints in gelatin, and then to use the phony prints to fool the reader protecting Joe's car or, worse still, Joe's place of work.

In theory, the results could be disastrous if Joe worked for NORAD or the Centers for Disease Control. In truth, neither NORAD nor the CDC nor, for that matter, any governmental or quasi-governmental installation is likely to abandon traditional security measures in favor of biometrics. They will instead view biometric devices as additive, something that augments rather than replaces. One can reasonably assume—or, at least, fervently hope—that those charged with protecting our most sensitive installations appreciate the need to secure them against the use of stolen biometric data by, for example, putting redundant systems in place.

Suppose instead that Joe plies a more mundane trade, working the day shift in an appliance warehouse guarded by a fingerprint scanner. Armed, so to speak, with Joe's gelatinous fingers, our clever thief could make his way into the warehouse of a dark night and make off with everything needed to redo his kitchen. If the thief somehow also managed to get hold of Joe's innovative Discover Card—which relies on a point-of-sale fingerprint scan to establish identity—the thief could then leave the warehouse for the nearest electronics superstore to load up on as many plasma screens as Joe's credit limit will bear. At this point, Joe's uniqueness—the very thing that should protect his identity—could prove his undoing.

This is so because the use of an indisputably unique biological or physiological characteristic to fix identity, if coupled with general confidence in the device responsible for assessing the characteristic, greatly decreases "deniability" should a false identification occur. Put another way, Joe's capacity to mount a successful challenge against the assertion that it was he in the warehouse and in the superstore bears an inverse relationship both to the uniqueness of the characteristic used to establish identity and to others' belief in the reliability of the device measuring that characteristic.

One of these—the uniqueness of Joe's prints—appears beyond question. The second—the confidence in the fingerprint reader's accuracy—is also likely to be quite high. While they might from time to time question the propriety of scientists' meddling in certain discrete areas—those involving, for example, heavy water and, later, stem cells—people, particularly Americans, hold strong, positive convictions about the ability and the utility of science and technology. If engineers claim that their fingerprint reader can distinguish real prints from fakes, people will tend to believe the engineers and to disbelieve someone claiming the machine has failed in its task. In the end, because Joe has fingerprints like no others, and because the devices used to read his prints stand at the technological cutting edge, Joe's claims that he was home watching television rather than out buying televisions will prove exceedingly difficult to maintain.

Representatives of the biometrics industry might respond that people have placed their confidence in biometrics technology well and that concerns about false identification are absurd, the product of paranoid fantasy. Biometrics technology, they insist, seeks to ensure and, more important, has the capacity to ensure that genuine and forged biometric data remain distinguishable. Deniability will suffer, to be sure, but only because it should suffer.

Those industry representatives would be right in one respect; the exploitation of stolen biometric data certainly is the stuff of fiction. The Greeks' so-called "new comedies" leaned heavily on disguise and assumed identity both to drive the plot and generally to amuse. The Roman playwright Plautus raised the stakes by relying on mistaken identity as the primary plotline in *The Twin Menaechmi*. Shakespeare—ever willing to find "inspiration" in the classics—revamped *The Twin Menaechmi* to come up with *The Comedy of Errors*. Moliere continued the tradition, almost invariably working at least one instance of stolen identity into each of his farces. The list goes on. From the *Decameron* to *Little Red Riding Hood*, disguise and stolen identity have proved a fictional mainstay.

It goes without saying that Burbank and Hollywood have built on this literary legacy, albeit in high-tech style. Stolen identity counts as all but a *sine qua non* of any episode of the *Mission: Impossible* television series. The same could be said for the Bond films, which depend on the forgery of everything from faces to fingerprints and from retinas to voiceprints. The numbers tell the story; in *Air Force One* and *X-Men II*, in *The Sixth Day* and *Seven* and *Ocean's Eleven*, in *Dracula 2000* and *2001: A Space Odyssey*, stolen and forged biometric markers drive the action.

The defenders of biometrics technology might turn these make-believe instances of stolen biometric data to their own ends, advance them as evidence that fears about false identification cannot be justified. They would have a point. Accepting that a wolf could play a grandmother so well as to dupe even the grandmother's kin or, for that matter, that Ethan Hunt—the lead character in the *Mission: Impossible* films—could use a mask and some manner of electronic throat patch to mimic perfectly almost anyone does require a certain suspension of disbelief. However, it would be a mistake to relegate concerns about forged and stolen biometric data to the realm of the fanciful. This is so because the stakes are high. Joe faces real prison time for the warehouse theft and will be responsible to pay for the televisions purchased at the superstore. Unflagging confidence in biometrics technology would leave his claims that he was at home unheard.

Even in these largely pre-biometrics days, effectively denying a false identification can prove difficult. Consider, for example, the story of Julia Twentyfive, a Las Vegas woman whose purse and, in turn, identity were stolen.[4] When the thief used information found in Ms. Twentyfive's effects to execute credit card scams, several felony warrants issued in Ms. Twentyfive's name.[5] Even after working for a year with authorities in two states in a bid to clear her name, Ms. Twentyfive found herself the subject of an ambush arrest, replete with a number of undercover officers and a film crew from the *Cops* television show.[6] She spent the next three days in jail.[7] After her release, Ms. Twentyfive spent yet another year and some $15,000 in attorneys' fees in an effort to set the record straight.[8] She ultimately found some repose only when an accomplice of the identity thief confirmed Ms. Twentyfive's innocence.[9]

Ms. Twentyfive's story is by no means unique. Thousands have found themselves devoting years and often some or all of their life savings in an effort to convince authorities and the credit bureaus that they did not engage in this or that transaction. Nor is Ms. Twentyfive's plight the most extreme. That honor likely goes to Gertrudis Rivera-Alicea, a resident of Allentown, Pennsylvania, whose Social Security number and birth certificate were stolen. The theft initially led to a fairly standard result: protracted confrontation with the Internal Revenue Service and the credit bureaus.[10] Matters took a decided turn toward the surreal, however, when Mr. Rivera-Alicea found himself at the business end of a paternity suit seeking support for an 11-year-old girl whose mother he had never met.[11] Fortunately, things were ultimately set right when the illegal alien who had stolen Mr. Rivera-Alicea's personal information admitted his wrongdoing.[12]

Imagine how a theft of biometric data might have compounded Ms. Twentyfive's and Mr. Rivera-Alicea's deniability problems. It is one thing to tell the police and credit bureaus about a snatched purse or stolen credit card. It would be quite another to explain that a thief must somehow have mimicked a retinal vessel pattern or a voice- or fingerprint so well as to fool an electronic scanner.

Again, this lack of deniability would be acceptable, even appropriate, if the device measuring a biometrics marker has a false acceptance rate at or very near zero. There can be little question, however, that at least some attempts to use forged or stolen data to fool measuring devices have succeeded.

Unfortunately, the evidence of these successes is of mixed quality. If the Web is any indication, latter-day Luddites and those with more specific misgivings about the effect of scientific advances on civil rights harbor deep suspicions about biometrics technology and seem willing to exaggerate its inefficacy as a matter of principle. That said, certain instances of triggered false positives appear valid, either because they have received widespread recognition or because they are described in careful and credible detail.

Research performed by Professor Tsutomu Matsumoto at Japan's Yokohama National University is, for example, quite well documented.[13] Working in materials ranging from gelatin to silicon, the professor fashioned fake fingers that fooled 11 fingerprint systems on a surprisingly consistent basis.[14] More striking, Professor Matsumoto forged digits not only with direct impressions from genuine fingers but also with latent prints lifted from objects.[15] This latter technique holds particular promise for would-be wrongdoers.[16] Rather than extracting a recognized print at knifepoint, they could find their key to the warehouse or research lab by simply rummaging through a dumpster.

Editors at Germany's *c't* magazine mounted a still broader attack on biometric scanners, going beyond fingerprint recognition systems to include iris-recognition and face-recognition devices.[17] The editors, who took obvious delight in their subversion, found themselves able to frustrate every system they tested. They repeatedly fooled so-called "capacitive" fingerprint scanners—which measure the relative stored electrical charge of a fingerprint's lines and troughs—by breathing on or placing a water-filled baggie on a scanner's sensor, which activated the device and caused it to reread the print left on the sensor by the last user.[18] Where no such residual print existed, the editors discovered they could dupe the readers by first lifting latent prints from glass or a compact disc onto adhesive film and then gently pressing the film onto the sensor. Optical and thermal recognition fingerprint scanners proved only slightly more problematic, forcing the editors to abandon warm breaths and adhesive tape in favor of fake silicon fingers.[19]

The *c't* editors also made short work of a facial-recognition device, finding that video footage of a registered user overwhelmed the device even at its highest security setting. In a cleverer twist, they confounded an iris-recognition device by combining high-resolution photos of registered users' irises with their own pupils.[20] The editors discovered after some trial and error that the device required a genuine pupil as a starting point for its analysis of the iris. Armed with that knowledge, they simply cut a small hole in their high-resolution photos, allowing the device to read their real pupils together with the photographed irises.[21]

These successes at replicating biological markers do not, to be sure, provide reason for abandoning a promising technology. They should none-theless serve as a caution. History teaches that thieves and pranksters have a tremendous capacity for mastering and then defeating protective tech-nologies. Confidence in biometrics technology should be tempered by this truth. Law enforcement personnel, regulators, credit bureaus, and credi-tors should entertain the possibility that the technology has, in a given instance, yielded a wrong answer. In short, some room for plausible deniability must remain.

Notes

1. *See* William Saito, *The Potential of Biometrics, at* http://www.g4tv.com/techtvvault/features/25668/The_Potential_of_Biometrics.html (Dec. 20, 2000).

2. *See id.*

3. Jonathan Kent, *Malaysia Car Thieves Steal Finger*, BBC News, *at* http://news.bbc. co.uk/2/hi/asia-pacific/4396831.stm (last visited Jan. 19, 2006).

4. *See* Alyson McCarthy, *Las Vegas Woman's Identity Theft Nightmare, at* http://www.klas-tv.com/global/story.asp?s=2066450&ClientType=Printable.

5. *See id.*

6. *See id.*

7. *See id.*

8. *See id.*

9. *See id.*

10. Phil Zinkewicz, *The Nightmare of Identity Theft, at* http://findarticles.com/p/articles/mi_qa3615/is_20036/ai_n9237962?cm_ven=YPI (June 2003).

11. *See id.*

12. *See id.*

13. *See* Tsutomu Matsumoto, et al., *Impact of Artificial "Gummy" Fingers on Fingerprint Systems, at* http://www.cryptome.org/gummy.htm.

14. *See id.*

15. *See id.*

16. *See id.*

17. *See* John Leyden, *Biometric Sensors Beaten Senseless in Tests, at* http://www.theregister.co.uk/2002/05/23/biometric_sensors_beaten_senseless/ (May 23, 2002).

18. *See id.*

19. *See id.*

20. *See id.*

21. *See id.*

Chapter 7

Biometric Authentication from a Legal Point of View— A European and German Perspective

*Astrid Albrecht**

I. Introduction

In addition to confidentiality, integrity, and availability, authenticity and thus the matching of an asserted identity with the actual one, is one of the main security goals related to information technology.[1] Of the three possibilities for personal authentication using knowledge, possession, or essence, biometric systems, based as they are on biometric recognition, are basically best suited to guarantee authenticity. While in the case of knowledge, an artificially generated code, such as a personal identification number (PIN) or a password, and in the case of possession, an element such as a card is only temporarily and indirectly allotted to a specific person by means of deliberate allocation, whereas essential characteristics such as physical attributes or behavioral are directly and, as a rule, permanently linked to a person. Such attributes cannot, on principle, be separated from a person, be it deliberately or involuntarily. By authentication using biometrics, electronic transactions can be reliably allotted to specific persons and can consequently create confidence in electronic legal transactions. Legal uncertainties, as well as unjustified allocations of liability

---

    *   Astrid Albrecht can be contacted at albrecht_astrid@web.de.

and attributions of the burden of evidence, can hence be better avoided than with other technical means.

This chapter offers a general overview of the advantages and disadvantages of biometrics mainly in the legal context and the important consideration of the technical performance in legal terms. This does not only cover privacy issues, which of course are a core element of the legal discussion, but also a general assessment of civil and civil procedure law aspects and possible contractual elements as trust building aspects in legal terms. This chapter takes a German view but also gives an overview of the recent European developments, mainly in the area of combating terrorism and the new generation of biometric-enhanced personal documents (especially travel documents). The view taken here is, naturally, not applicable directly to other legal systems. However, some of the thoughts expressed herein are supposed to tackle the legal discussion on biometrics, which is about to catch up with the technical promotion of biometric authentication. The content and references are mainly based on spring 2005.

II. The Different Aspects of Biometrics

The identification of people from biometric characteristics is a promising approach to the authentication of individuals. In addition to or instead of conventional methods such as PIN/password and card or other tokens, biometrics has the advantage that physical characteristics are directly tied to particular individuals, unlike knowledge and possession. Biometric methods are based on the premise that certain physical characteristics can be assigned to one particular individual, who is identified on the basis of his individuality. Physical characteristics are directly linked to the body of the person and therefore do not have to be artificially assigned. In contrast to characteristics that merely reference the person, they are tied to the person directly and not just by inference. Physical characteristics cannot be lost either. Owners of particular characteristics do not have to remember them because they carry them around with them at all times. Generally these characteristics cannot be kept secret either; on the contrary, many of the physical characteristics used for biometric identification are obvious. Finally, unlike the PIN/password and card or other token, biometric characteristics cannot be transferred or passed on.

The aim of biometric recognition is always to ascertain the identity of a person (identification) or to confirm or reject the identity claimed (verification). There is a requirement to distinguish authorized persons from unauthorized persons and/or to verify a claimed identity. Along with confidentiality, integrity, and availability, authenticity is one of the paramount security objectives in the information technology context. If, therefore, the physical characteristic is correctly assigned to a person, then use of that characteristic can provide assurance that the person present really is the person who he is assumed to be or who he purports to be.

The revolutionary technology of biometrics thus offers new points of departure for increasing security. However promising biometrics appears to be, as with all new technologies, it is necessary to examine carefully the advantages and disadvantages of its methods. From the point of view of IT security, which is, for instance, the primary concern of the German Federal Office for Information Security (BSI),[2] the significant aspects are recognition performance and security. On top of this, it is necessary to consider both the legal principles and the social aspects, such as the acceptance and ease of use of a biometric application, since IT security never stands alone by itself.

III. Interoperability by Standardization

Standardization can provide interoperable solutions for which the user is not forced to depend exclusively on the know-how of specific manufacturers and vendors. At the international level, the International Standardisation Organisation (ISO) established Sub-Committee SC 37 (ISO/IEC ITC1/SC37) as the main international body to work on biometric standardization by the end of 2002.[3] SC 37 consists of six working groups and the respective shadow committees on national level. The working groups are considering standards for common vocabulary (WG 1), for technical interfaces (WG 2), for interchange formats for templates and images (WG 3), for application profiles architecture (WG 4), for testing and reporting accuracy (WG 5), and for legal and social aspects of biometrics (WG 6).

SC 37 WG 6 is developing a technical report on "Cross-Jurisdictional and Societal Aspects of Implementation of Biometric Technologies," which is concerned with legal, social, cultural, and ethical issues relating to biometric methods. The aim is to achieve an internationally harmonized and practicable assessment of biometrics, which above all gives developers and users of biometrics guiding principles for legally binding and socially acceptable applications that go beyond the purely technical point of view. This challenge should not be underestimated, as significant differences exist not only in the cultures of different countries of the world but also, and especially, in national legal frameworks. This task can only be mastered if, first, the committee, with its international composition, perceives and respects the existing differences, and second, an endeavour is made, despite all the differences, to find the smallest common denominator with which quite basic values can remain safeguarded in all societies and cultures when using biometric methods.

On the German level, within the Deutsches Institut für Normung e.V. (DIN), the German Institute for Standardization, the Normenausschuss Informationstechnik (Information Technology Standards Committee (NI)) is concerned with biometric issues dealt with by SC 37 on a national level. DIN is a registered association founded in 1917 with its head office in Berlin.[4] Since 1975, it has been recognized by the German government as the national standards body and represents German interests at the international and European levels. The area of work of the NI includes the development of standards in the field of Information Technology, with the aim to improve the productivity/efficiency and quality of IT systems, to enhance the security of IT systems and data, to support the portability of software programs, to ensure the interoperability of IT products and systems, to unify software development tools and environment, and to ensure the designing of ergonomic user interfaces.[5]

DIN NI 37[6] deals with all aspects listed above and contributes to all SC 37 Working Groups with German comments and delegations for the SC 37 meetings on ISO level. This includes work on CBEFF—Common Biometric Exchange Formats Framework—and BioAPI—the Biometric Application Programming Interface. NI 37 consists of 34 regular members. The chairman of NI 37 is a German representative of Fraunhofer Institut für Graphische Datenverarbeitung; his representative is a senior official of the German BSI.

IV. Process of Biometric Identification

To appreciate the technical security and data security aspects of biometric methods, a certain basic understanding of the way in which they function is necessary. The basic principle of biometric recognition is the same under all the systems.[7] All biometric systems contain the following elements, regardless of their technological design, which will often be highly individual:

- personalization or registration of the user in the system (enrollment),
- capture of biometrically relevant characteristics of a person,
- creation of data sets (templates), and
- comparison between the characteristics presented and those previously stored (matching).

Both on the first occasion on which data is collected for the reference data set and also on the later occasion of checking, biometric characteristics are recorded using sensors such as cameras, microphones, keyboard, pressure pad for signature recognition, or fingerprint sensors.

To record a person in a biometric system, first an image of the original characteristic is generated and recorded. This is the raw data. Using an algorithm that is generally manufacturer-specific, this original is transformed into a data set, known as the template. This contains the data set extracted from the recorded data. On the other hand, for a pure comparison of images, no template is generated, but the original image is stored as reference image and compared with a new original image.

During the matching process, a comparison is then carried out between the template stored and the data set that is created upon subsequent presentation of the characteristic to the biometric system. If the two agree, the device reports that the person presenting the biometric characteristic is recognized. Recording, evaluation, and comparison of biometric features are naturally vulnerable to measurement errors, as the features used alter during the course of time. This can be due to natural changes related to the aging process or to external factors such as injury or illness. Then there are external changes, such as changes of hairstyle (haircut, beard), the wearing of spectacles or contact lenses, and changes in cosmetics. Moreover, the characteristic is never presented to the system by the user in exactly the same way. The result is that two digital images of a single biometric characteristics can never be identical. The actual decision as to whether there is a match or not depends much more on the matching score and on pre-set parameters. These give rise to tolerance ranges, in which biometric data is evaluated by the system as either a match or non-match. Thus, the biometric characteristics are tested not for identity, but only for "sufficient similarity." If the comparison values fall outside the applicable tolerance range, then an error occurs—either a false rejection or a false acceptance. The probabilities of these errors occurring are referred to as the false rejection rate (FRR) and the false acceptance rate (FAR) respectively.

V. Data Security through Biometrics

In the information society, biometrics can make life more secure and easier[8] and, as an aid to enhancing data security, it is an attractive prospect.[9] Data security is the complete set of organizational and technical (not legal) controls and measures with which unauthorized access to personal data is prevented and the integrity and availability of the data and the technical facilities used to process it are maintained.[10] Data security thus expresses a philosophy of data reduction and data economy. Technical and organizational measures, as defined under Section 9 German Federal Data Protection Act (BDSG),[11] are aimed at safeguarding the implementation of the law and the requirements of the Annex to Section 9, which include access control (physical entry, access to system, and access to data), transmission control, input control, availability control, and guarantee of limitation to a specific purpose.

With regard to data security, the close link between biometric characteristics and a person is to be welcomed, as this differentiates the systems significantly from authentication systems, which are based solely on possession and knowledge. Genuine person verification is possible. The condition, however, is that the correct attribution of the identity of a person to the reference data must actually be assured.

Above all, given the known unreliability of knowledge elements, a system based on biometrics can therefore bring considerable gains in data security—for example, for access authorization in networks, access to networks, and the secure storage of valuable data.[12] Thus, the principles stated in the Annex to Section 9 of the BDSG regarding the requirement for the security of personal data can in principle be better assured if biometric systems are used. This applies especially with regard to access control (paragraphs 1-3 of the Annex). Here, a biometric system can serve to guarantee the security of data requiring protection in other applications, and hence the necessary authenticity.

VI. Security of Biometric Systems

On the other hand, however—the other side of the coin, as it were—where biometric systems are employed, if physical characteristics of a user should ever be compromised, that is, misused or forged—it is basically not possible to replace or revoke them, as would be possible with a secret number or password.[13] Moreover, the pool of characteristics available for use in such a system is very limited, compared with knowledge-based codes.[14] The degree of data security that can be achieved with biometric methods thus depends on the security of the systems themselves.[15]

The security of biometric systems depends largely on the protection of the reference data and the comparison mechanisms.[16] Three aspects are particularly important here. First of all, the reference data must actually originate from the characteristics of the person to whom it is attributed. Secondly, the integrity of this data—its genuineness—must be assured both during enrollment and forever afterwards. Finally, the input data that the sensors extract from the biometric features must not be intercepted and replayed, nor must it be possible to reproduce it without the involvement of the user. One of the special features of biometric data that needs to be considered here is the fact that biometric recognition can only produce a probability statement; it cannot deliver 100 percent certainty. The error rates that occur with all biometric systems (false acceptance of unauthorized persons and false rejection of authorized persons) have to be considered in every application where it is a matter of gaining the expected security on the one hand and providing the necessary legal protection of personality on the other.

VII. Security Infrastructures for Biometric Systems

As well as the technical conditions, trust in biometric systems requires support through organizational and legal measures up to a point. Trust in the legal sense presumes that the person who is to be trusted will also behave as such trust would warrant in the future.[17] On the other hand, from an organizational point of view, generally recognized evaluation criteria and their confirmation by independent bodies within the framework of a certification process can contribute toward the creation of trust. It is becoming increasingly essential that independent, competent third parties be able to check and verify the technical security of IT systems. Again, evaluation and certification are also very important with biometric systems, not least so that their quality and trustworthiness can be assessed.[18]

With regard to evaluation, the aims of IT security criteria are to be a yardstick for the development of secure, trusted systems. The objective evaluation of these systems by a neutral and competent entity, compared with manufacturer declarations alone, will guarantee a minimum level of trust and enable the user to select a suitable IT security product for his or her specific needs. Certification means the confirmation of adherence to evaluation criteria by independent bodies. The Common Criteria for Information Technology Security Evaluation[19] is an internationally approved ISO-standard.[20] In Germany, the Federal Office for Information Security, BSI, is the agency that evaluates and certifies IT-security products.[21] Such "Trusted Third Parties" can thus confirm, for instance, a claimed performance of a biometric system so that the deployer does not need to rely only on the vendor's claims, which are market-driven.

One widely acknowledged example of neutral evaluation criteria for biometrics is the "Best Practices in Testing and Reporting Performance of Biometric Devices" of the UK-based and internationally manned Biometric Working Group.[22]

Also worthy of mention in this context is the German Association TeleTrusT e.V.,[23] which the Biometric Working Group (WG 6) has published under others as a Bioethics Guide for the objective comparison and assessment of biometric systems.[24] The catalogue includes technical aspects as well as legal and privacy aspects of biometrics, which can be crucial when implementing a biometric system in a real world application.

VIII. Implications of Data Protection to Biometrics

The ambivalence of biometric technology becomes especially evident in the area of data protection. On the one hand, biometrics allow for authentication that is secure and in compliance with data protection rules. On the other hand, biometrics involves new risks especially regarding the protection of users' personal rights.[25] Biometric methods thereby create a new kind of risk potential[26] and increase the vulnerability of the individual.[27]

The following example illustrates this in the context of personal rights. In a specific data protection context, biometric methods can be used, for example, to fight activities known as identity theft. Identity theft is increasingly becoming a problem, especially in the United States. This term designates activities in which somebody uses personal data assigned to another, for example, to perform financial transactions.[28] Most commonly in the area of electronic payment transactions, one finds theft of a name, Social Security number, or credit card data.[29] If it were possible to better protect this data against theft by using biometrics, unauthorized use would likely no longer be as easy and, in any case, it would not suffice to merely obtain code numbers assigned to individuals.

Looking at it from another angle, however, the intrinsic quality to the person that characterizes biometrics also constitutes the drawback.[30] Other risks not present when traditional methods are used arise from the "rootedness" of electronic transactions enabled by biometric methods, similar to a signature written in the person's own hand.[31] Regarding our earlier example of identity theft, new risks may arise when a person's biometric data is obtained and used without authorization for purposes of identity theft. It follows that both the chances and the risks of biometric methods hinge on the characteristic ability of biometrics to identify a person by the use of intrinsic physical characteristics.[32]

A. Legal Framework for Data Protection in the European Union

With regard to the crucial data security when using biometric data demonstrated above, all those mechanisms exist in order to realize the data protection provisions defined by law. The German data protection law is based, as is all European data protection regulations, on European law. Therefore, this section describes the fundamental basics of European data protection law before getting into the details of the German national provisions.

1. Legal Basis in Europe

The legal basis in Europe for treating personal data includes the following three main legal frameworks.[33]

First, the Treaty on the European Union, Title I-Common Provisions-Article F states:

The Union shall respect the national identities of its Member States, whose systems of government are founded on the principles of democracy. The Union shall respect fundamental rights, as guaranteed by the European Convention for the Protection of Human Rights and Fundamental Freedoms signed in Rome on 4 November 1950 and as they result from the constitutional traditions common to the Member States, as general principles of Community law. The Union shall provide itself with the means necessary to attain its objectives and carry through its policies.[34]

Second, the European Convention for the Protection of Human Rights and Fundamental Freedoms, Article 8, states:

Everyone has the right to respect for his private and family life, his home and his correspondence. There shall be no interference by a public authority with the exercise of this right except such as is in accordance with the law and is necessary in a democratic society in the interests of national security, public safety or the economic well-being of the country, for the prevention of disorder or crime, for the protection of health or morals, or for the protection of the rights and freedoms of others.[35]

Last, but not least, the European Union Data Protection Directive, which is also the basis for necessary amendments of all European Data Protection laws in the European Member States,[36] gives the main framework for data protection in the EU. According to Art. 1 (1) of the Directive, the object of this Directive is the following: "In accordance with this Directive, Member States shall protect the fundamental rights and freedoms of natural persons and in particular their right to privacy with respect to the processing of personal data."[37] Recital (2) states: "Whereas data-processing systems are designed to serve man; whereas they must, whatever the nationality or residence of natural persons, respect their fundamental rights and freedoms, notably the right to privacy, and contribute to economic and social progress, trade expansion and the well-being of individuals."[38]

2. Core Values of the EC Data Protection Directive

The core values of the directive can be summarized as follows:

- Reduction of the processing of personal data to the unavoidable extent.
- Maintenance of the highest transparency possible.
- Maximization of efficiency of institutional and individual control of processing of personal data—for example, specific rights of the data subject with regard to his/her personal data.

The Directive can also be considered on the basis of eight core principles regarding the processing of personal data:

- Fair and lawful processing (legal limitation or consent required).
- Specified, explicit and legitimate purpose (the finality principle).
- Respect for the right of the data subject.
- Data kept in a form that permits identification for no longer than necessary for the purposes for which the data were collected.
- Proportionality—adequate, relevant, and not excessive.
- Accurate and up-to-date.
- Appropriate technical and organizational measures against unauthorized use or unlawful processing.
- Data may only be transferred to those countries that ensure an adequate level of protection for the personal data.

Whereas the national data protection laws need to be compliant with the main principles of the EU-Directive, specific approaches can also be realized for certain aspects. Thus, the national legislation differs in some aspects of data protection, which either mean a stronger or a weaker protection of personal data.

3. Article 29 Data Protection Working Party

A Working Party was set up in order to advise on data protection and privacy—the Article 29 Data Protection Working Party (Art. 29 WP).[39]

This Working Party was set up under Article 29 of Directive 95/46/EC. It is an independent European advisory body on data protection and privacy. Art. 29 of the Directive 95/46/EC states that a Working Party on the Protection of Individuals with regard to the Processing of Personal Data, is set up with advisory status and that it is to act independently. It defines how the Working Party is composed, how decisions are to be made, and other formal regulations. Its tasks are described in Article 30 of Directive 95/46/EC and Article 15 of Directive 2002/58/EC[40] and include making recommendations on all matters relating to the protection of persons with regard to the processing of personal data in the European Community (Art. 30 Nr. 3).[41]

Meanwhile, Art. 29 WP has issued several working documents and statements related to biometrics and the current developments on biometric enhanced passports and identity cards.

The first publicly available statement of the Working Party to biometrics was "Working Document on Biometrics," WP 80 as of August 1, 2003.[42] The purpose of the document was to contribute to the effective and homogenous application of the national provisions on data protection.

WP 80 expresses a general concern regarding the increasing use of biometrics as personal data and the need of protection of those data in an appropriate manner in accordance to the EU Data Protection Directive in almost all cases. The document includes a short description of what biometrics are and how they work in general, followed by a summary of some main principles of the EU Data Protection directive in terms of biometrics. Understood as the core of the protection by Directive 95/46/EC, the principle of proportionality is emphasized herewith.

Option No. 7/2004 on the inclusion of biometric elements in residence permits and visas, taking account of the establishment of the European information system on visas (VIS) as of August 11, 2004,[43] followed as WP 96 with regard to specific aspects of the implementation of biometrics in personal documents.

WP 96 deals with the decision of the Thessaloniki European Council of June 19 and June 20, 2003, which had confirmed that "a coherent approach is needed in the EU on biometric identifiers or biometric data, which would result in harmonized solutions for documents for third country nationals, EU citizens' passports, and information systems VIS and SIS II" and invited the European Commission "to prepare the appropriate proposals, starting with visas." The European Commission, in due course, submitted a draft

Council Regulation, amending Regulations 1683/95 and 1030/2002, which laid down a uniform format for visas and for residence permits for third-country nationals, respectively. The proposal provided for the mandatory storage of digital photographs of the face and fingerprints within two and three years, respectively, of their adoption; a contactless chip would serve as the storage medium. On September 30, 2005, Opinion 3/2005 followed as WP 112 on Implementating the Council Regulation (EC) W. 2252/2004 of December 13, 2004, on standards for security features and biometrics in passports and travel documents issued by member states.[44]

B. Data Protection in Germany Relating to Biometrics

Data protection in Germany follows the EU Data Protection Directive directly. Within a given timeframe, all EU Member States were obliged to transpose the Directive into national law. The German Federal Data Protecting Act has been adopted with the new legislation of 2001. The following section will give an introduction and an overview of main German data protection principles according to German law, and some possible implications related to biometric data.[45]

1. Right to Informational Self-Determination

Data protection law in Germany is understood as the special right of legal protection of personality. Section 1 (1) BDSG states the individual is to be protected against suffering any impairment of his right to privacy through the handling of his personal data—that is, more than just protection of personal data. The protection objective is the right to informational self-determination, which is defined as the power of individuals to basically decide whether their personal data should be divulged and used. This is to be guaranteed through procedural framework conditions for the collection of data.

The German Federal Constitutional Court stated, for the first time, in a basic decision on population census in 1983, the right to informational self-determination, which has been, since then, anchored in the German Constitution,[46] Arts. 2 (1) and 1 (1) (right to general freedom to act and human dignity) and is widely acknowledged as an unwritten basic human right according to the German Constitution:[47]

Under modern conditions of data processing, free development of the personality presupposes protection of the individual against open-ended collection, storage, use and disclosure of his personal data. This protection is therefore subsumed in the basic rights contained in Arts. 2 (1) and 1 (1) German constitution. To this extent the fundamental right guarantees the power of individuals to basically decide for them whether their personal data should be divulged and used.[48]

This right has several crucial implications for processing personal data in accordance with German law. A special justification is always required where the decision-making power of the data subject could be restricted. Thus, a general prohibition of processing is provided that can only be lifted by law, other regulation, or the consent of the individual, 4 (1) BDSG ("right to withhold permission").

2. Biometric Data as Personal Data According to German Data Protection Law

The right to informational self-determination is the stated aim of data protection in Germany. It includes the power of individuals to basically decide whether their personal data should be divulged and used. This must be assured through procedural framework conditions governing the collection of data. Section 1 (I) BDSG states that the aim of data protection in Germany is to protect the individual from suffering any impairment of his or her right to privacy through the handling of his personal data. It follows, then, that one central aspect of the assessment of biometric data in the context of data protection law is the link between that data and the data subject. The scope of protection of informational self-determination is always affected and, indeed, only when personal data is involved. If this is the case, then one must assume infringement on the right protected under constitutional law whereby special justification is required, either through specific provisions in the law or else through the consent of the data subject. Section 3 (I) BDSG defines personal data as "any information concerning the personal or material circumstances of an identified or identifiable individual."

In this definition, the legislature has not further specified what possibilities of re-individualization and exactly what additional information should be assumed when checking identifiability. Whether the biometric data does actually contain a direct link to the data subject depends critically on the technical implementation. As each technical implementation is inevitably different, whether the specific form of processing and storage chosen allows a link to be made to a particular person or to an identifiable person can be contested in every individual case. As far as the link to the data subject that is contained in the data is concerned, the usability and possibility of using this data is critical; these in turn depend on the purpose for which the data is collected and the specific possibilities associated with IT of processing and linking it.

Art. 2 (a) EC Data Protection Directive provides a point of reference as to when a person is deemed to be identifiable: "[A] person is regarded as identifiable when that person can be identified directly or indirectly, in particular through attribution to an identifying number or to one or more specific elements which are an expression of that person's physical, physiological, mental, financial, cultural or social identity."

The Federal Constitutional Court made it very clear in the 1983 decision mentioned that, "Given the possibilities of automatic data processing, there is no such thing as insignificant data any more." This means that in any circumstance personal data is to be secured with respect to the right of informational self-determination.

There are different stages of biometric data that have to be taken into account when considering biometric data as possible personal data—raw data and templates.[49] Whereas raw data always points to a natural person, the situation with templates can be slightly different. Templates are encrypted raw data and can be stored in different ways, that will then determine their scope of use. They can be stored centrally, decentrally, or locally (so called self-authentication) and using specific approaches—template-free methods and encrypted biometrics.

However, direct identification of a person using his or her physical characteristics is basically always possible as biometric characteristics always relate to one particular person only. At the same time, it is not critical which (technical) tools enable identification. A link to the data subject may exist for one organization, but might not be possible for another. Hence, in order to define biometric data as personal data, we need to take a relative view. The data can be anonymous for one party and personal for another. So there is no such thing as abstract personal data but, in each case, this has to be defined by the application in question.

3. Purpose Limitation

The principle of limitation to a specific purpose (also referred to as finality principle) acquires special significance with biometric data. According to German data protection law, personal data may only be processed when a given purpose is defined before processing. Changing the use of the data that is no longer appropriate to the original purpose is basically prohibited under Sections 14 and 28 (I 2) BDSG. Due to the lifelong association between biometric data and the individual concerned, it is harder to prevent the data acquiring an independent life of its own than in the case where only indirectly personal data is involved. In the latter case, destruction of the data also destroys the link to the data subject. To this extent, further use of the biometric data beyond the intended use could become more problematic than in the case of other personal data so securing the biometric data is even more crucial. In view of the qualitative transformation of the data that occurs with biometric methods, it is therefore important to define precisely what specific purpose the data serves. The more specific the purpose, the more difficult it becomes to gain new information through links. Thus, the lifelong linking of the biometric characteristic to the data subject can be, with respect to this aspect, a disadvantage of biometrics in general whereas, as stated above, in terms of data security, this lifelong linking is desirable.

With regard to this principle, the Federal Constitutional Court stated another basic principle of the right to informational self-determination in the context of the general control of the subject over his or her personal data:

> However, individual self-determination presupposes, even under the conditions of modern information processing technologies, that the individual retains the freedom to make decisions as to which actions to take or not take, including the possibility of actually behaving in accordance with these decisions. Anyone who cannot keep track with adequate certainty of what information concerning him is known in particular areas of his social environment and of who may not care what possible communication partners know about him, can be significantly hindered in his freedom to plan or to take decisions based on his own self-determination. A social

order and a legal system in which citizens can no longer know who knows what when and on what occasion about them cannot be reconciled with the right to informational self-determination. Anyone who is uncertain whether divergent behaviors are taken note of at any time and stored permanently as information, used or divulged to third parties, will try to escape notice by avoiding such behaviors.[50]

4. Materiality Theory

The legislature has a duty to make all the material decisions in basic normative areas (materiality theory). The increasing importance of information in society brings with it special protection obligations on the part of the legislature, which, among other things, relate to the creation of organizational and legal framework conditions that allow the individual to participate in the IT society without any restriction or loss of his or her capacity to act. One current example is the Anti-Terrorism Act,[51] passed by the Deutscher Bundestag (the German Parliament), which went into effect on January 9, 2002, and contains provisions amending a large number of security laws in line with the new threat situation. In the amendments to the acts governing passports and identity cards, it is stated that for further details another federal law is required.[52]

In some areas where the European legislator is setting up rules by regulations and directives that are not to be transposed into national law but are entering into force immediately in each member state, it might be argued that the materiality theory is no longer valid or at least does not gain as much importance as it was intended to in the past. However, such assessments must be made on a case-by-case basis and cannot be judged in general.

5. Proportionality Principle

To grant a reasonable guarantee of the legal protection of personality, the data protection legislation stipulates that in every application one should reference the principle of proportionality and ask whether any legitimate interests of the user can outweigh the data subject's private interests and whether the use of biometric methods is actually necessary for the defined purpose and is also suitable for achieving the desired goal. When one examines the issue from this angle, the actual recognition performance and the demonstrated security of the biometric system become particularly important. The more vulnerable to misuse a system is, the less legal justification there is for encroaching on the right of informational self-determination.

6. Prohibition of Discrimination

The technical peculiarity of biometric systems, namely the fact that the decision as to whether a person is recognized or rejected depends on a threshold value, could also be problematic when one considers the possibility of discrimination against data subjects. False rejections of authorized persons could lead to false alarms, which could have serious consequences for the data subject, depending on the application context. False acceptances, on the other hand, could lead to counterfeiting of the data attributed to an individual and to this extent impair his informational self-determination. The right to informational self-determination protects the individual against any stigmatization and any burden of justification resulting from this. Moreover, unjustified trust in biometric technology can lead to cases of discrimination, when the individual is excluded from certain benefits due to a possible incorrect biometric decision. In many applications it might be necessary and sensible to provide an additional manual check of a biometric decision.

7. Biometric Data as Sensitive Data

With regard to a special protection needs of biometric data, again the principle of proportionality gains importance. According to European and German law, among personal data there can also be a category of "sensitive data." Section 3 (9) BDSG, Art. 8 EC Data Protection Directive and Art. 6 EC Data Protection Convention define such data as data ". . . from which [a person's] racial and ethnic origin, political opinions, religious or philosophical convictions, union membership, health or sex life can be inferred."

With this data there is a special danger of discriminatory use. Questioning this category for biometric data opens the discussion on the sense of such an abstract category in general. An abstract assignment to categories without sufficient consideration of the specific context in which the data is used is problematic. Also, again according to the Federal Constitutional Court, there is no such thing as insignificant data any more, as discussed above. Hence it seems to make more sense to always weigh the conflicting interests. Oriented to the specific processing operation, there are the meriting interests of the data subject versus the justifiable interests of the responsible organisation that have to be taken into account. Basically, where a predominant general interest exists, one can come up against the boundaries of exercise of the right to informational self-determination.

8. Biometrics and the Concept of Privacy-Enhancing Technologies

Finally, biometric methods are also of interest in connection with the concept of privacy enhancing technologies (PET). This is a coherent set of information and communication technology measures that protect the private sector by eliminating or reducing personal data or preventing the unnecessary or undesirable processing of personal data, without the information system losing its functional capability. PET reflects a philosophy of data reduction and data economy that covers a whole system of technical measures. According to Section 3a of the BDSG, data reduction and data economy are to be achieved by data processing systems collecting, processing, and using no personal data or as little personal data as possible.

The principle of PET can apply to biometrics from two standpoints. First, the implementation and application of biometrics has to follow a correct privacy regime in order to be privacy enhancing. Second, biometrics can be a privacy enhancing method. For instance, the European Initiative on Privacy Standardization in Europe speaks in its final report of biometrics as a tool of privacy enhancing technologies.[53] Also, Art. 29 WP emphasizes the use of privacy enhancing technologies with regard to biometrics.[54] The main question according to the concept of PET is whether or not "identity" is necessary for each of the processes of the conventional information system. In most cases it is not necessary to know the user's identity in order to grant privileges. Yet there are some situations in which the user must reveal his or her identity to allow verification.

Whereas biometric systems, if suitably designed, could thus be turned into an implementation aid and an important component for PET, this concept can itself be applied to the technical design of biometric systems. This includes measures such as rendering data anonymous, using aliases, and encrypting data.

IX. Biometrics and Border Control

As part of the effort to combat international terrorism, the authorities are interested in the use of biometric technology to improve identity verification at the various stages of checking new arrivals and residence entitlement.[55] This is currently the most topical application for biometrics. The primary basis for such endeavors is the Anti-Terrorism Act,[56] passed by the Deutscher Bundestag, which went into effect on January 9, 2002, and contains provisions amending a large number of security laws in line with the new threat situation. The subject has also gained in importance at the international level. The U.S. Congress has passed a legislative package aimed at combating terrorism that, among other things, includes major changes to the visa waiver program. This requires that participating states, including Germany, incorporate biometric characteristics into their travel documents by October 2005. In accordance with the aims of the American "Enhanced Border Security and Visa Entry Reform Act" of May 14, 2002, based on the U.S. Patriot Act, for the introduction of biometrics into the travel documents of visa waiver states, the International Civil Aviation Organization (ICAO, a UN body) is examining recommendations to extend travel documents to include biometric characteristics. The ICAO specifies the use of facial recognition as the biometric characteristic for global interoperability, but leaves the door open to other optional characteristics, such as fingerprints or iris scans.

A. European Development

As a result of these international developments, the European Council Regulation, (EC) 2252/2004, on standards for security features and biometrics in passports and travel documents issued by member states entered into force on December 13, 2004.[57] According to this, passports and travel documents must be provided with a storage medium (radio-frequency chip) that contains a facial image, and fingerprints are to be added in interoperable formats.

The main components of radio-frequency chip (RF-chip), based on radio frequency technology, are a tag (e.g. a microchip) and a reader. The tag consists of an electronic circuit that stores data and an antenna which communicates the data via radio waves. The reader possesses an antenna and a demodulator that translates the incoming analogue information from the radio link into digital data. The digital information can then be processed by a computer.[58]

Whereas the facial image has to be added to EU passports by mid-2006, fingerprints do not have to be added until 2008. Art. 6 of the Directive specifies down that member states shall apply this regulation with regards to facial images no later than 18 months and, with regard to fingerprints, no later than 36 months following the adoption of the technical measures referred to in Art. 2 (see below).

The core principles of the European Council Regulation, (EC) 2252/2004, on standards for security features and biometrics in passports and travel documents contain, among others, the following regulations.

First, passports and travel documents issued by member states shall comply with a minimum security standard. These standards are set out in the annex of the Directive and include the material that is to be choosen with regard to the paper of the document, details of the bibliographical data page, possible printing techniques, protection means against copying, and the issuing technique.

According to Art. 1 Nr. 2, passports and travel documents shall include a storage medium containing a facial image and fingerprints in interoperable formats. It is mandatory that the storage medium has sufficient capacity and capability to guarantee the integrity, the authenticity, and the confidentiality of the data.

Nr. 3 makes clear that the regulations set up in this Directive do not apply to identity cards issued by member states to their nationals. Hence, it remains the member states' decision as to whether or not—and how—to incorporate biometric data into national identity cards.

Art. 2 is the core of the Directive regarding further technical specifications of the travel documents. It is stated that, in addition to what is already regulated in Art. 1, additional specifications shall be established in detail for the following aspects:

- additional security features and requirements, including enhanced anti-forgery, ant-counterfeiting, and anti-falsification standards;

- technical specifications for the storage medium of the biometric features and their security, including prevention of unauthorized access; and
- requirements for quality and common standards for the facial image and the fingerprints.

Art. 4 Nr. 1 grants the right of the individuals to verify the personal data contained in the passport or travel document and, "where appropriate" to ask for rectification or erasure. Nr. 3 sets up the principle of purpose limitation and states that biometric features in passports and travel documents shall only be used for verifying the authenticity of the document and the identity of the holder by means of directly available comparable features when the passports or other travel document are required to be produced by law.

As of February 28, 2005, the decision of the European Commission on the Specifications of EU-Passports based on and amending the European Council Regulation (EC) 2252/2004 just mentioned here, especially Art. 2, entered into force.[59] This document describes solutions for chip enabled EU passports and is based on international standards, especially ISO standards[60] and ICAO recommendations on Machine Readable Travel Documents. It specifies:

- specifications for biometric identifiers: face and finger;
- storage medium (chip);
- logical data structure on the chip;
- specifications for the security of the digitally stored data on the chip;
- conformity assessment of chip and applications; and
- RF compatibility with other electronic travel documents

For the primary biometric-face, the standard compliance is given with ICAO NTWG, Biometrics Deployment of Machine Readable Travel Documents, Technical Report, Version 2.0, 5 May 2004, and ISO/IEC FCD 19794-5: Biometric Data Interchange Formats–Part 5: Face Image Data. For the secondary biometric–fingerprints, the standard compliance is given with ICAO NTWG, Biometric Deployment of Machine Readable Travel Documents, Technical Report, Version 2.0, 5 May 2004, ISO/IEC FCD 19794-4: Biometric Data Interchange Formats–Part 4: Finger Image Data, and -2 Biometric Data Interchange Formats–Part 2: Finger Minutiae Data, and ANSI/NIST-ITL 1-2000 Standard "Data Format for the Interchange of Fingerprint, Facial, Scarmark & Tattoo (SMT) Information"; FBI: Wavelet Scalar Quantization (WSQ). Last but not least, for the storage medium (RF-chip architecture), the standard compliance is given to ICAO NTWG, Biometrics Deployment of Machine Readable Travel Documents, Version 2.0, 5 May 2004, ISO/IEC FDIS 14443, Identification cards—contactless integrated circuit(s) cards—proximity cards, and ICAO NTWG, Use of Contactless Integrated Circuits In Machine Readable Travel Documents, Technical Report, Version 3.1, 16 April 2003.[61]

It is important to note that the United Kingdom and Ireland have not taken part in the adoption of this measure whereas it is obligatory for 19 EU member states, including Germany, France, Greece, Italy, The Netherlands, and Spain.

B. Biometrics with Regard to Personal Documents in Germany

In Germany, all new passports issued after November 2005 contain an RF-chip and a digital facial image. As a second step, the addition of digital fingerprints is planned, too. Old passports will remain valid when issued before that date. In the context of personal documents, next to the described latest developments on the EU level that have directly influenced Germany, the Anti-Terrorism Act of January 9, 2002, introduced essential changes that affect the German passport and the German identity card and established the legal base for the decision mentioned above.

1. Amendments to the Acts Governing Passports and Identity Cards

As well as the existing photograph and signature, German passports and identity cards are, according to current legislation, can contain other biometric features in digitized form. The amendments made to the acts governing passports and identity cards make it possible to incorporate other biometric data relating to a person's fingers, hands, or face, in addition to his or her photograph and signature.

Art. 7 and 8 of the Anti-Terrorism Act contain the amendments of the act governing passports (Passgesetz, PassG) and the act governing identity cards (Personalausweisgesetz, PAuswG).[62]

Section 4 (III, 1) and 2 PassG, respectively, and Section 1 (IV, 1, II) PAuswG state: "The passport . . . may contain, next to photograph and handwritten signature, additional biometrics from fingers or hands or faces of the passport holder . . . Photograph, handwritten signature, and the additional biometrics may be implemented also in encrypted manner."

Section 4 (IV, 1) PassG and Section 1 (V, 1) PAuswG further state, "The used biometric characteristics, their details and the concrete implementation of biometrics and other data in encrypted manner as well as the way of storing and usage are to be regulated by further federal law." Hence, a special federal law will specify the types of biometric data that can be used, the details thereof, and the incorporation of data and information in encoded form, as well as the way in which the data is stored and the ways in which it can be processed and used. This solution was thought to facilitate the evolution of personal identity documents, while bearing security requirements in mind. Since 2002, when the discussed changes of the respective laws were laid down, the European legislator has now decided, as mentioned before, that it is mandatory for Germany to deploy the changes for face and fingerprint, at least for the EU passport, according to the defined time schedule set out in the respective EU Directive.

In the interpretation of these acts, it was stated that the additional biometric data will enhance computer-aided identification of persons on the basis of identification papers. The reliability of identifying a person merely by comparing the person's looks to his or her photograph is seen to be dependent on the observer's subjective perception, and it can be impaired by a large number of factors, such as the quality of the photograph, the natural aging process, and changes in a person's hairstyle, beard, etc. Incorporating additional biometric data is therefore seen as the prerequisite for enhancing the possibilities of identifying a person on the basis of his or her identification papers. By introducing the possibility of integrating biometric data into passports or identity cards, data that is used in encoded form in security procedures is understood to make it possible to establish unequivocally whether the identity of the person holding the documents matches the identity of the person being checked, and will do so using a computer-aided system.

2. Data Protection and Security Provisions with Biometric Enhanced Personal Documents

The amendments of the acts governing passports and identity cards include specific provisions for data protection and security.

(i) Legal Provisions

Section 4 (IV, 2) PassG / Section 1 (V, 2) PAuswG states, "There will be no Germany (federal)-wide database established." As has been the case so far in Germany, no data records will be stored centrally in the future. This is to avoid any potential misuse of personal data and the risk of person tracking with regard to centralized storage.

Section 16 (IV) PassG / Section 3 (V) PAuswG says, "The encrypted biometrics and other data in passport/ID-card may only be used for checking the authenticity of the document and for identity check of the holder." This regulation is designed to provide the needed limitation to a specific purpose. Thus, it is not allowed to use or process the biometric data out of this scope.

Section 16 (VI, 2) PassG / Section 3 (V, 2) PAuswG state, "By request the passport or ID-card agency must provide the holder of the document with information on the content of the encrypted biometrics and other data." This aims to give the most possible transparency for the individual whose biometric data are used for identification purposes. Regarding the right of informational self-determination mentioned above, it is important for the individual to have the possibility and the right to know what data concerning himself or herself are stored.

(ii) Technical Means

Technical mechanisms are provided to protect the biometric data stored on the passport by means of the following procedure:[63]

Basic access control (BAC) requires the optical reading of the machine readable zone as a precondition to access of the personal data stored in the chip. Once finished with the technical protocol, the reader is identified as the reader which has optical access to the passport. At the same time, there is a common secret being created which secures the following communication between reader and passport against unauthorized eavesdropping by means of appropriate encryption-secure messaging.

As an alternative with even increased security, the reader itself has to be authorized against the RF-chip in order to communicate with it. This step is based on Extended Access Control (EAC). This technical protocol requires that the reader has its own cryptographic keys and an additional key certificate to be verified by the RF-chip. In this certificate the rights of the reader have to be defined exactly. Within the technical protocol, there is again a common secret to be created that then secures the following communication between reader and RF-chip against unauthorized eavesdropping by means of appropriate and, even stronger, encryption compared to BAC.

X. Authenticity in Legal Transactions Through Biometric Actions

The following sections are based on the author's doctoral thesis (see note 45).

In traditional, paperbound legal transactions, the written form serves as evidence of the acceptance of obligations and provides assurance of the rights and obligations of the parties.[65] Generally, no formal requirements apply in private legal transactions.[66] It is assumed that mandatory compliance with statutory form would severely impede legal transactions.[66] Hence, the basic absence of formal requirements should take due account of the realities of modern commerce.[67] As long as a particular form is not required either by law or by agreement between the parties, legal transactions can be effectively concluded without observing any formalities. If, however, a requirement to observe a particular form is laid down, then failure to observe that form invalidates the legal transaction (Section 125, German Civil Code, BGB).[68] To facilitate the conclusion of transactions where compliance with statutory form is not mandatory, the sole reason for the validity of the intended legal consequences is the outwardly recognizable intention of the parties to be bound by the transaction.[69] Nevertheless, numerous agreements are concluded in writing in private law that, above all, are based on traditional trust in the uniqueness of the written signature as a biometric action, and a high value is attached to this in the sense of a social state of trust.[70]

A. Functions of the Written Form and Importance of Handwritten Signature

Insofar as the law requires the written form, the formal requirements in civil law require that legally relevant declarations are signed by the author manually with a personal signature or with a mark authenticated by a notary public (Section 126 I German Civil Code). The signing of a declaration always presupposes that both the declaration and the signature occur in the same document: the signature must appear physically at the end of the text.[71] The personal signature required by law should make it possible to identify the author.[72] The author is the originator of the declaration stated in writing—that is, the person who gives the declaration in his own name or as the representative of another person acting on his own responsibility.[73] The writing must be individual and unique, exhibit characteristic features that adequately identify the person signing, and make it difficult to imitate—if not completely excluding such a possibility.[74] The signature is only deemed to be personal if it is produced with the author's own hand.[75] A handwritten signature therefore requires that the motion of writing is conveyed as such onto the document.[76] These requirements serve above all to establish authenticity, that is, to establish the connection between the declaration embodied in the document and its originator. In legal transactions, the handwritten signature is thus the typical characteristic that is used to establish the originator of a document and that person's intention to put into circulation the declaration committed to paper.[77]

Adherence to the written form is related to certain security purposes:

> The need to observe a form elicits a businesslike state of mind amongst the participants arouses legal awareness, calls for circumspect reflections and ensures the seriousness of the decision made. The observed form further makes clear the legal character of the action, serves, rather like the impression of a coin, as a stamp of the executed legal will and eliminates all doubt as to the finality of the legal act. Finally, the observed form provides proof of the existence and content of the legal transaction for all time; it also leads to a reduction in or to the shortening and simplification of the processes.[78]

It was only through the development of the signature, the first evidence
of which was provided by the cuneiform script of the Sumerians around
3000 B.C., and the much later invention of the mechanical letterpress by
Gutenberg in the 15th century, that it became possible to record thoughts
and intentions independently of purely verbal delivery and, as a result, to
make them provable and capable of being communicated to third parties
independently of the individual powers of recall of the person concerned.
The first "informatization" occurred around 500 B.C. in Greece with the
transition from purely verbal communication to a form of communication
that was also—and critically—in writing.[79] It was the culture of writing that
made it possible to communicate binding declarations of intent over a distance and to establish legal relations between absent parties. Perpetuation on
paper made "abstract" or depersonified transactions possible. In the event of
a dispute, the parties ceased to be dependent solely on their memories, which
could be mistaken.[80] The signature below an agreement is thus of importance not simply in the legal sense in case of dispute, but it is also the expression of a social state of trust between the parties.[81]

The conferring of privileges following the provision of a handwritten
signature in conventional legal transactions is based on the following functions:[82]

- An identity function. The author of the declaration allows himself to
  be identified; the unique tie to the person signing is established through
  the signature or the mark authenticated by notary public.[83]
- A conclusion function that signals the finality of the declaration (as
  compared with the simple draft) and the deliberate nature of the act of
  giving it. Signature and mark are placed at the physical end of the text
  and state that the intention of the signer which precedes it is correct
  and complete.
- An associated warning function: In particular, this protects the signer
  from excessive haste and makes clear to him that the document is no
  longer a draft but a legally relevant declaration with binding consequences.
- An evidence function that can also be viewed as going hand-in-hand
  with the conclusion and warning functions.[84] The signature has the
  effect of creating documentary evidence of the finality of the legal

transaction executed, the conclusion of the contract is clearly distinct from pure negotiations, the content of the contract is fixed and clarified, and proving what has been agreed to becomes easier.

- Providing evidence of receipt of a written declaration. It is necessary to ensure that declarations sent to the recipient have actually reached that person, and the written signature can serve as evidence of such receipt.[85]
- Written documentation of the receipt of papers by the recipient. This is particularly relevant where a fixed period of time is initiated upon receipt of the document which, if missed, will have adverse legal consequences for the recipient.[86]

Thus, the written form has, above all, a preventive purpose—namely, it can offer certainty as to the obligation or entitlement to a service "for every case."

B. Peculiarities of Electronic Documents

Electronic transactions entail electronic documents, as opposed to written documents. The term "electronic document" serves here to characterize the medium that contains the declaration. The electronically signed document is technically a signed file that normally contains an electronic declaration of intent.[87] A declaration of intent basically expresses a desire for certain legal consequences—that is, an intention that aims at establishing a private legal relationship or making changes to the content of or terminating an existing one.[88] An electronic declaration of intent is distinguished by the use of electronic means with respect to both its creation and its transmission and receipt.

Where information and communication technology is used to conclude legally binding transactions and where electronic documents are used, the physical evidence of the declaration of thoughts discussed above in relation to the written form is basically absent.[89] Neither written documents nor the handwritten signature are transmitted in the original. Instead, electronic data is incorporeal and, hence, transient. Moreover, instead of using his own hand, the author of an electronic document makes use of technical means, such as a smartcard that he inserts in a card reader, and normally he will initiate and execute the signing process by entering a PIN and pressing some keys. In principle, the security functions inherent in the written form,[90] especially the link between the text and its author who tradition-ally makes it binding with his handwritten signature, are lacking here.

As electronic media become more and more widely used, the question arises whether and to what extent the evidence-providing elements linked with the written form can be satisfied and whether and to what extent the security requirements and expectations of the parties can be met. It must be assumed that the parties to an electronic legal transaction have a similar need to establish certainty as to authenticity and authorship.[91]

To some extent, the needs of modern legal transactions are already being met, both by case law and by legislation. Mechanisms have been developed to avoid obstacles to a use of new technologies in this context. The use of new technologies has been discussed primarily in relation to form requirements for procedural briefs. Procedural briefs are those that directly perform or are designed to perform a party's procedural act.[92] This includes briefs that commence or terminate proceedings.[93] Preparatory briefs, on the other hand, have no immediate procedural effects and merely serve the purpose of preparing and facilitating the court hearing (Section 129, German Civil Procedure Code).[94] No special form requirements ap-ply to procedural briefs.

More recently, the requirement for briefs filed in court proceedings to be signed in the legal counsel's own hand has been linked to its function by the courts, which have clarified that this rule is solely intended to en-sure that the document should show with sufficient reliability both the contents of the declaration to be made and the person from which it origi-nates. Further, it must be certain that the document is not a draft but has been communicated to the court with the appropriate person's knowledge and intent.[95] The courts have also clarified that with this form requirement the only appropriate test can be a reasonable degree of strictness.[96]

While relaxing the written form requirement is quite appropriate in communications with the courts—in closed user groups—it would not be appropriate in legal transactions among private parties in the general pub-lic. The main argument against dropping the requirement of a signature in the declaring person's own hand is that a signature has a very different degree of relevance in communications with courts, where it only exists as a "should" rule that moreover is being interpreted in accordance with its function and with a view to the specific requirements of procedural law. Insofar as form requirements are to be relaxed, it will be necessary to look at the specific technical solution used to verify accurately whether the functions of the written form can be satisfied. One will further have to keep in mind that new communication technologies also make it difficult to adequately prove receipt of an electronic document.[97] In principle, one will have to stick with a strict interpretation of the written form requirement. To conclude, electronic documents do not satisfy the written form requirement pursuant to Section 126 BGB even taking into account the discussion relating to procedural briefs.[98]

This means that in this area biometric authentication, methods are only of minor relevance. In communications with courts and subject to the limitations noted above, the requirement of a signature in the counsel's own hand remains unapplied where new communication technologies are used. In these cases, there is no absolute need for using biometric acts. One should, however, keep in mind that in the use of faxes, the need for protection against falsification is not generally satisfied as fully as would be appropriate in view of the general need for security, even in communications with courts. In this regard, the courts appear to lean toward an uncritical faith in technology. In some cases a reference to a relevant operator's manual is considered enough and it is assumed that "manipulations and interference by third parties can be largely prevented by technical precautions."[99] Other courts refer to "sufficient other indications" that may be used similarly to a signature to ascertain the identity of the person making the declaration and the intent of sending off the declaration, thereby satisfying the need for legal certainty.[100] In a very few cases, certain lower courts point out the risk of falsification when new technologies are used.[101]

If in the future it were to become possible to use adequately secure biometric systems (for example in preparing a computer fax in which the signature would be replaced by automatic signature recognition), then certainty of evidence could be ensured even in communications to courts. At least in communications to courts, it would be easy to perform recognition of electronic documents prepared in such a way and using such technologies.

C. Certainty of Evidence Obtained Through a Signature in the Declaring Person's Own Hand

In traditional legal transactions, the signature in the declaring person's own hand is of special importance. This is reflected in the certainty of evidence rules. The parties' expectation to be able to prove rights and obligations in legal transactions indicates the special need for certainty to keep specific evidence available for arrangements made and covers all the relevant functions of the written form. The need for functional equivalence in electronic documents is a consequence of a similar need for certainty regarding electronic legal transactions. The centuries-old tradition of the written form and the social trust based on it cannot easily be mimicked here. As manipulation is easier and information is generally more fleeting in electronic media, special security mechanisms are needed for electronic legal transactions in order to achieve a state equivalent to legal transactions bound to traditional paper. At least in legal transactions between private parties, media that lend themselves to unauthorized use comparatively easily, such as faxes, are inadequate for this purpose. In communications to courts, as a result of the revision of Section 130 no. 6 ZPO, such media are now correctly held to be adequate; here case law previously developed by the courts has been codified. In communications between courts and legal counsel, the resulting security deficiencies are more acceptable than in legal transactions between private parties. Here communication takes place within a closed user group. Within this user group, one may assume that on one side the courts will generally be trustworthy and on the other legal counsel and/or parties have no increased need of protection. The sole purpose of the written form is to ensure that procedural briefs have been intentionally filed by the proper legal counsel or by the party itself. Conversely, in legal transactions between private parties, the written form serves the added special purposes of warning and protecting the parties involved. These purposes, however, cannot be served by electronic media. This means that in these cases one cannot drop the requirement of a signature in the declaring person's own hand.

However, consensus exists that there is a need for electronic communication to be facilitated even in legal transactions between private parties. Therefore Section 126a BGB introduced the electronic form. Here the mechanism of the qualified electronic signature is intended to satisfy the need for security, elsewhere satisfied by a signature in the declaring person's own hand. While the security infrastructure established by the signature provisions in law and regulation is basically capable of satisfying expectations as to the certainty of the electronic form, one essential purpose of the signature in the declaring person's own hand cannot be served as long as the signature is operated and released solely through a PIN mechanism. This cannot ensure the authorship of a signature created in this way. This means that, even where a qualified electronic signature is used to comply with the electronic form under Section 126a BGB, a substantial authenticity gap remains because a PIN can never verify the identity of the person acting. On the contrary, the signature can be created by any third party because it is assigned only indirectly to the authorized signature's owner through the PIN and signature card. It cannot be ensured in this way that the signature is actually created by the authorized person, which would be necessary to achieve a functional equivalence of the electronic form to the written form. This means that when a PIN is used, the certainty of evidence necessary to satisfy the parties' expectations is not provided. The term "electronic signature" is actually a misnomer because when technical media are used, the immediate and inseparable link between the writing tool, the hand, and the acting person that alone characterizes a signature is not present.

According to the view taken here, the traditional written form cannot be adequately reproduced in legal transactions among the general public as long as merely a PIN authentication mechanism is used to create and issue a legally binding electronic declaration. Such a mechanism, which is merely assigned to a person and is linked to the acting person merely derivatively and indirectly, cannot substitute the traditional signature in the acting person's own hand. Finally, with regard to the parties' needs for security that should be satisfied even in electronic legal transactions, the situation can be improved by using sufficiently secure biometric authentication mechanisms. This is especially true if one were to use biometric systems that required an active and conscious act on the part of the person and did not permit an unintentional or unknowing release of, for example, the signature mechanism, it would be possible to adequately reproduce the characteristics of a signature written in the person's own hand, thereby adequately serving the purposes of the written form.

D. Biometric Methods and Legal Proof

The use of technical methods can be acknowledged in hearing evidence even without applying statutory evidence evaluation rules. For some methods, empirical rules have been developed on the basis of general life experience under which rendering proof may be facilitated (applying the rules of prima facie proof) in certain circumstances for the party having the burden of proof. These empirical rules derive primarily from a recognition that the technical method in question is sufficiently secure.[102] Generally, the court determines the truth of the facts submitted by the parties by hearing evidence during the proceedings. Under general civil procedural rules on the burden of proof, the party asserting a claim must submit and prove the facts on which the claim rests.[103] If proof cannot be rendered, the consequences are on the party that has the burden of proof. Burden of proof is not a duty of the party to render proof, but merely states who will suffer if proof cannot be rendered. Burden of proof is "the risk of one party to lose a case because such party has been unable to prove the facts on which its case rests."[104] The burden of proof is allocated in accordance with an unwritten general principle: The claimant has the burden of proof for facts supporting the claim, the defendant has the burden of proof for the facts negating or invalidating the claim or hindering its enforcement.[105] The burden of proof rules apply if, despite having explored all the evidence available under procedural law, the court still cannot reach a conclusion on whether a disputed and relevant fact is true or untrue (non liquet). Generally speaking, there are risk-allocation rules underlying the allocation of the burden of proof.[106] They derive from value judgements and are open to interpretation and to be used mutatis mutandis. They can be added to or modified by the courts.[107]

In certain special circumstances, rendering proof can be facilitated for the party having the burden of proof. One of the rules applied to facilitate rendering of proof in a majority of cases of unauthorized uses of debit cards is prima facie proof. Prima facie proof may be made available in cases in which, taking into account all undisputed and certain individual facts and special circumstances of the case and in accordance with general life experience, a typical unfolding of events can be identified for the fact to be proven.[108] This, according to general life experience, must indicate a specific cause and must appear as so usual and common that the special individual circumstances lose relevance.[109] This rule of prima facie evidence has not been laid down in any law in Germany and its legal root remains disputed.[110] However, it has been applied by the courts as customary law for many years.[111]

The evidentiary certainty of any technical method and, therefore, the legally binding quality of electronic transactions made using such method depend essentially on the actual security of the methods and technical components used.[112] The evaluation of how secure a technical method is, especially by the court hearing evidence, substantially affects "fairness of evidence." If the court is generally convinced that the method used is secure, then this can be a disadvantage to the party to which the proof is being applied, if the security is merely presumed.

E. Facilitated Proof in the Use of Biometrics

The evidentiary value of an electronic transaction can be maximized if one or more adequately secure biometric systems are being used as protection against unauthorized access and to issue an electronic legal declaration. If the biometric system used is sufficiently secure, then one may assume that the degree of certainty of evidence can be enhanced, not only in electronic signature methods, but wherever proving the authorship of an electronic transaction may become relevant. Both in a court's evaluation of evidence and in the application of the evidence evaluation rules examined above regarding the use of qualified electronic signature methods, the level of trust based on a signature written in a person's own hand and the resulting certainty of evidence can be mimicked in electronic transactions more adequately by using appropriately designed biometric systems than through an authentication mechanism that is merely assigned to the person, such as a PIN.

Given the current state of biometric technology, however, one must doubt whether this means that in cases where biometric systems have been used, it will be justified to facilitate the rendering of proof beyond the degree of privilege already available as courts freely evaluate evidence.

Since rules facilitating proof modify the normal allocation of risks of proof and since the allocation of these risks under general burden of proof rules corresponds to a general principle of freedom, each such modification must be reviewed as to whether it is in compliance with the German

Constitution (GG).[113] One might consider using the rule of law principle in Art. 20 III GG as a test for whether a shifting of the burden of proof should be permitted.[114] According to a consistent line of cases from the German Constitutional Court (BVerfG), Art. 103 I GG demands fair proceedings;[115] this serves to protect a party against being stampeded, neglected, left to cope on its own, or assigned inequitable burdens in proceedings.[116] This means that specific types of evidence may be privileged only if they are reliable and their actual meaningfulness is foreseeable for the parties.[117] Fictitious proof unsupported by actually reliable evidence, on the other hand, is not permitted.[118] Moreover, it remains to be considered that Art. 3 I GG also demands procedural equality of weapons for the parties, to ensure that equal treatment of the parties extends to the application of procedural rules.[119]

It follows from the above that statutory facilitation of proof rules may be introduced only under a number of conditions and only if the technical method in question and the application systems satisfy a large number of technical and organizational conditions that actually justify the legislator's assumption.[120] The examples regarding the use of the insufficiently secure PIN authentication mechanism to prove authenticity of debit card transactions and of electronic signatures have shown that if the security of a certain technical method is merely assumed and naïve faith in technology is unfounded, there will be a lack of necessary legal certainty. This uncertainty may render unjust results when evidence is heard, resulting in disadvantages to one of the parties. The test for the value of a technical method in evaluating evidence can therefore only be how secure the method has been proved to be. If biometric systems are to be used as suitable evidence, they will have to provide adequate levels of technical security in order to be assigned a high value as means of evidence. A number of requirements for the security of biometric systems have been enumerated.[121] It should always be kept in mind that there is no absolute security and that such absolute security also cannot be determined to be a characteristic of any technical system. Instead, the security of a technical system is the result of a value judgement made on the system in question.[122]

For cases in which certain technical methods have been used, rendering proof may be facilitated by way of a statutory presumption of security, such as in Section 15 I 5 SigG for qualified electronic signatures with offeror-accreditation; this then creates a presumption that the method used is secure. This has the benefit that the person who applied the presumed secure method will be supported by this presumption of security in court.[123] The presumption facilitates proof of an electronic legal declaration because the person who has the burden of proof is no longer required to prove that the method is secure. In this way the burden of proof is allocated justly and in a manner friendly to persons applying them. The decisive benefit is that the information that the person having the burden of proof would have to provide under general burden of proof rules has already been provided by competent bodies and is included by way of the statutory presumption of security. This saves the parties the effort of explaining and proving the details of the security of the method used in each and every case. This saves effort for both users and courts, which may now refer to the statutory evidence evaluation rule.

One has to take into account specifically that a biometric system can only perform to near 100 percent recognition.[124] It is therefore difficult to make reliable statements on the recognition quality of a specific system at a defined point in time.[125] It follows that a statutory evidence evaluation rule for biometric systems at this point in time might even result in increased risks if it subsequently turns out that the system privileged with the evidence evaluation rule was more vulnerable to damage than had been assumed. In this case, it might be impossible or very difficult for the party to whom the evidence evaluation rule is applied to prove the presence of system errors and manipulations. An evidence evaluation rule privileging biometric systems might possibly deprive the party against whom evidence is being brought of any opportunity to prove that his or her person had not in fact been authenticated. This would be all the more true if, similar to Section 292a ZPO, the mere use of the method identified in the provision would result in an assumption that the authorized person had been authenticated correctly. This, however, would be justified only if in view of the (technical and practical) experience with biometric methods in general or with certain biometric systems it could be ruled out with reasonable certainty that an unauthorized third party could pretend to be the authorized user through unauthorized use and manipulation. In view of the enumerated possibilities of defeating a biometric system by unauthorized means and of the lack of security infrastructure, such a conclusion is not currently justified.[126] In the end, such an unjustified facilitation of proof could cause considerable damage (to trust) both micro- and macro-economically, depending on the extent to which biometric systems were actually used.[127]

Moreover, in the context of electronic transactions, biometric systems are normally used purely as access protection to control access to protected data or mechanisms. Again it cannot be ruled out that the authorized user may have obtained access to the protected area via biometric recognition. In other words, he or she has opened it with proper authorization, and subsequently a third party (whether noticed or unnoticed) has, say, helped himself or herself to the protected data. The procedural defense raised by the party who has become the victim of such an intrusion must not be entirely eliminated by an evidence evaluation rule that grants privilege to the party applying the biometric system.[128] In the context of electronic signatures, a special difficulty might arise for the owner of the signature to try to establish reasonable doubt within the meaning of Section 292a ZPO about having intentionally communicated the signature, if biometric systems were supported by an unjustified assumption of a high degree of security. But in practice, even without a statutory presumption of security, a court could simply assume that when biometrics have been used, a prima facie proof would be more difficult to rebut than if a traditional PIN had been used. In this way, unjustified naïve faith in technology could result in a higher degree of legal uncertainty than that which already exists in cases in which traditional authentication methods are used. For the matter of primary and secondary attribution here in question, it is essential to determine what proven and not merely assumed degree of security biometric methods will have when used in electronic transactions.

Quite generally it makes no sense for the law to anticipate a technical development as long as the law remains able to respond flexibly to technical innovations. The court's freedom in evaluating evidence is a achievement of the enlightenment that is worth maintaining even in the face of new technologies.[129] Moreover, evidence evaluation rules are characterized by an inflexibility that some authors regard as out of place in a technically evolving field.[130] The law of evidence generally is not an appropriate tool for forcing innovative technologies. It should not be adapted to an untested technology. Instead, the technology should be given a chance to prove itself. This means that the only legal methods of evaluating evidence that are risk adequate are those that take positively into account for court decisions the available, confirmed experience and that are able to respond flexibly to new experience and discoveries. In civil procedure law this is ensured by the court's freedom in evaluating evidence under Section 286 ZPO. Confirmed experience, which will develop over time in dealing with the biometric products in question, can subsequently (once adequate experience is available) be used by the courts through allowing prima facie proof,[131] although again the consequences of unjustified naive faith in technology must be considered to avoid unjust shifting of the burden of proof.[132]

It follows that without the inflexible concept of evidence evaluation rules, the development of the further details by the courts can be oriented more closely with practical experience and requirements, and that the development of empirical rules on the use of biometric systems will lead to case law that remains continually open to critical discussion and that, where unusual circumstances are present, it will allow for decisions diverging from the empirical rules in the interest of giving justice in the individual case. In hearing evidence on technically elaborate modern technologies in particular, courts will have to critically evaluate the means of evidence offered, where necessary by involving an expert or risk-making judgments that may be unappealable formally, but will be erroneous in their content.[133]

The application of the rule of the court's freedom in evaluating evidence to the use of biometric systems is not hindered by the fact that, due to its possible lack of knowledge about the new technology and its basic technical principles, a court may be unable to evaluate the evidence correctly. In evaluating evidence, science and logic, psychology and human experience prevail, which naturally means that the "judge, as such, is not an expert."[134] This is true for practically every case in which evidence is heard. Very often in hearing evidence, courts have to deal with matters the backgrounds of which they may not easily understand. The existence of court divisions with special competencies for one thing and the involvement of experts for another mitigates potential problems here. The expert assists the court in evaluating the facts submitted by stating personal opinions, conclusions, and hypotheses on the basis of his expert knowledge.[135] In doing so, he or she is bound to the facts provided by the court because, under the principle of immediate evidence, Section 355 ZPO, finding the facts is the court's responsibility alone. Pursuant to Section 144 ZPO, the judge must decide whether, in evaluating relevant facts, he or she has reached the limits of his or her knowledge and of the options for learning more available in his or her capacity as a judge and will need to consult an expert. The court may leave finding the facts to the expert only in exceptional circumstances, namely when the expert's special knowledge is required even at the fact-finding stage.[136] In these cases, the expert will be an expert witness under Section 414 ZPO. It must be assumed that this happens in the majority of cases involving an evaluation of complex technical sets of facts when the court is unable to find the facts by itself.[137] It follows that doubts in the courts' ability to evaluate evidence correctly cannot justify an introduction of statutory rules limiting the court's freedom in evaluating evidence.[138]

If, as one suggest here, the courts remain free in evaluating evidence, then the party invoking the use of a secure biometric system will suffer no disadvantage. On the contrary, such party will have all options for providing evidence open in proving the actual and manipulation-free authentication of the authorized person. Where a system that is provably adequately secure against manipulation has been used, the court (which will likely and out of necessity be supported by an expert) would not be able to argue without providing points of attack for an appeal that in the case to be judged an unauthorized third party has pretended to be the authorized person.

When it becomes possible for certain biometric systems to be certified by independent bodies on the basis of accepted security criteria as being sufficiently reliable and secure against manipulation to provide a proof of such nature,[139] it might be justified to limit the courts' freedom in evaluating evidence of this sort. Until then, in reaching decisions, courts will have to deal with the security of the methods used and will develop case law privileging such methods after assembling appropriate general life experience—that is, after collecting appropriate proof.

This means that if no evidence evaluation rule is introduced, this will not hinder the use of adequately secure biometric systems. Instead, a court's free evaluation of the evidence provided will reflect the actual degree of security adequately and in accordance with the state of the art in this technology as it evolves. This will ensure that risks and burdens of proof are allocated justly and in a manner friendly to the persons applying these systems and thereby complies with the principle of fair proceedings.

XI. Trust-Building Options in the Law

Beside implementing reasonable technical requirements, designing appropriate liability provisions is one important aspect in building trust in the use of biometric methods. The aspect of objective verifiability of previously made promises of security should be supplemented with corresponding and enforceable provisions on the legal side: "Trust requires the presence of facts indicating that the person who is to be trusted will act in such a way in the future as the trusting person expects in view of having invested trust."[140] Security requirements for biometric systems that are merely postulated or even set down in the law would remain ineffective if noncompliance with them were to remain unsanctioned.

A. Standard Terms of Business: Liability Clauses and Duties of Skill and Care

Liability can be dealt with not only in statutory provisions but also in contracts. Businesses often use pre-worded contract terms that, as a rule, diverge from the statutory provisions. The law of standard terms of business, laid down in the German Standard Terms of Business Regulations, contains tests for whether such terms have become part of a contract and whether clauses are invalid and entitles third parties (such as consumer protection associations) in certain circumstances to file a court claim aimed at obtaining a judgment holding a clause invalid. Generally, what is needed is clear and understandable wording. In other words, the consumer must be able to find and familiarize himself or herself with certain mandatory information without difficulty.

The risk of malfunctions of the biometric system that prevent the user's access to the protected area may not be generally shifted to the user in standard terms of business. This applies both to problems resulting from a change of the physical characteristic and for technical access and system malfunctions. Besides the need to deal with the physical characteristic in a socially adequate way and the resulting permitted degree of risk, this is also due to unavoidable and uncontrollable changes of the physical characteristic and to the way that the frequency of errors is handled within the biometric system's scope of tolerances, which is in the operator's sphere.

An exclusion of liability for system malfunctions caused by the operator is also probably not permitted in standard terms of business. Biometric systems are vulnerable to unauthorized use and malfunctions. In addition to intentional manipulations by attacks on the system and general technical problems, a certain number of potential erroneous acceptances must be expected to occur. The user is able neither to affect nor to control these technical aspects. They are in the operator's sphere alone. Moreover, the operator has made unauthorized uses of the application possible by introducing the system in the first place, thereby creating the risk of unauthorized use. Because the falsification risk in a biometric system can be affected and mitigated by the operator's appropriate safeguarding measures, which are outside the user's control, such risks cannot be shifted to the user by standard terms of business. Whether and in what way the operator of a biometric system will be permitted to shift parts of the risk of unauthorized uses to the customer depends essentially on the degree of security that can reasonably be achieved and on the safeguarding measures that are taken.[141]

This means that one must look more critically at attempts to place responsibility for specific actions on the user in the context of a use of biometric systems than in the case of codes that are generated artificially and are merely assigned to the authorized user indirectly. While it is critical to keep PINs and passwords secret, because of arising questions of unreasonable hardship, a duty of care relating to a physical characteristic appears as outright strange and therefore cannot be used to impose legal burdens on a user through clauses in the operator's terms of business. We have seen that many of the actions theoretically required from users to ensure the smooth working of a biometric system are simply impossible or at least, in most cases, would constitute unreasonable hardships. In the final analysis, the rules on prima facie evidence that work with a violation (presumed applying common sense) of responsibilities relating to a use free of malfunctions and unauthorized uses of a certain biometric system, cannot apply here.

XII. Employee Data Protection

In Germany, as in other European countries, there are specific regulations on the data protection of employees in the workplace that lead to particular legal conditions for the operator of a biometric system in the workplace.

The legal protection of personality is set especially under Section 75 (2) Employees' Representation Act (BetrVerfG). Here, a joint duty of employer and works council is established in order to protect and promote the free development of the personality of employees working in the business. This also contains the guarantee of informational self-determination of the employee affected by a biometric system.

The regulations in this area in particular require the balancing of conflicting interests. This raises, at any rate, the question of whether or not a pressing operational interest in the use of a biometric system can be founded on an objective basis as well as a special security interest. The basis for this is the general data protection rule that excessive data collection is prohibited, so that for each data obtaining a reasonable purpose is to be given.

Furthermore, the covert monitoring of employees is prohibited in the absence of a justified interest in certain information (e.g. reason to suspect involvement in theft). This is due to the ultima ratio principle: only in the case where means other than surveillance do not serve the defined purpose is it allowed to covertly monitor employees. This is also not allowed during proceedings for protection against wrongful dismissal either. In this case, the use of proof is prohibited as well. German courts have also decided that there is no such thing as a dismissal on the basis of automated (based on biometrics) decisions that is based on section 6a (1) BDSG; such decisions (like dismissal) that have a serious impact on a person's life must not be based on the outputs of electronic software and systems only.

The co-determination of the works council over technical monitoring devices is regulated in Section 87 (1) No. 6 Employees' Representation Act and affects directly the implementation and introduction of a biometric system. It states that a technical device exists in any case if the behavior or performance of the employee is brought to human notice at least partly by this device. According to this section, monitoring is a process under which performance or behavior-relevant data on employees is collected and then analyzed in some form. For this purpose, the objective suitability for monitoring (i.e., the potential suitability for monitoring) is sufficient, because otherwise the judgement as to whether or not the process fell within the definition would be left to the subjective interpretation of the employer.

The use of a biometric system to control entry through a gate or door or as an access system, time recording, or attendance-monitoring system is included as a rule.

This applies at any rate where a link to the data subject can be established, as the employer is then in a position to determine (and evaluate) the presence of the employee concerned.

If access is to be denied only to unauthorized persons, then no direct link to the data subject, i.e., no link to the individual employee, is necessary; rather, in this case the biometric system is merely a substitute for a key, and Section 87 (1) No. 6 of the Employees' Representation Act does not apply. However, it is important to note that if logs of the recognition process are maintained for error analysis (FAR/FRR), identification of the person will normally be possible.

In general, Section 87 (1) No. 6 will apply to any biometric system at the workplace. This means that an existing working council is to be included in the decision making process of whether or not, and if yes, in what manner, a biometric system will be installed. The German association TeleTrusT e.V. has developed a guideline for a factory agreement on the use of biometrics in the workplace (available at www.teletrust.de).

Recent judicial decision of the German Federal Labour Court on co-determination over the secondment of staff to customer site with biometric access control system has established another important rule for biometrics at the working place. The court said:

> The works council has to be involved where an employer instructs its employees to undergo [identity checking] at a customer's site by a biometric access control system that is in use in that establishment. The instruction concerns the behavior at work of customer service staff on secondment and, under Section 87 Para. 1 No. 1 Employees' Representation Act, therefore requires the involvement of the works council. Moreover, the case involves the use of a technical monitoring device requiring the involvement of the works council under Section 87 Para. 1 No. 6 Employees' Representation Act. The fact that the access control system has been set up at a

customer's site does not preclude the co-determination right of the works councils. Although the employer has no direct influence on circumstances at the customer's premises, nevertheless it is giving the seconded staff instructions that require the involvement of the works council. An agreement must therefore be concluded between the employer and the works council as to whether and in what manner the employer's staff are to be subject to access control at a third party site. When drawing up the contract with the customer, the employer must ensure that the agreements made with the works councils are implemented. The rights of the individual employees involved remain unaffected by this.[142]

Therefore, even if the employee is sent to a foreign company to fulfill a task, the working council of the sending company needs to be involved in the decision of the use of a biometric system.

XIII. Outlook

Biometrics can improve the legal situation with regard to accountability and evidentiary rules in electronic transactions. If the biometric system in place had the approved level of reliability and had ideally an approved performance—adequate error rates also in terms of misuse—the legal liability of transactions where biometrics are used can be increased. This includes an objective assessment of the biometric application and requires that neither the operator nor the user have unjustified trust in the system based on unproved facts. If these conditions were met, courts were able to judge cases with an appropriate acknowledgment of the technology used.

To grant a reasonable guarantee of the legal protection of personality, the data protection legislation stipulates that in every application one should reference the principle of proportionality and ask whether any legitimate interests of the user can outweigh the data subject's private interests and whether the use of biometric methods is actually necessary for the defined purpose and is also suitable for achieving the desired goal. When one examines the issue from this angle, the actual recognition performance and the demonstrated security of the biometric system become particularly important. The more vulnerable to misuse a system is, the less legal justification there is for encroaching on the right of informational self-determination.

Finally, the design of the technology will have a significant impact on the way the statutory data protection requirements are implemented. As advances in technology accelerate, data protection becomes increasingly dependent on that technology. Data protection through technology is therefore rightly seen as a "landmark" of the EC Data Protection Directive and the German Federal Data Protection Act. Finally, advances in technology can be harnessed so as to incorporate the concept of PET into technical data protection. However, technical requirements are purely accessory precautions and, to this extent, they cannot have a primary regulative function, but only a purely instrumental one. Hence, although the resolute implementation of data protection-friendly concepts offers the possibility of supporting data protection in its regulatory function, this cannot mean dispensing with normative concepts in data protection.

Notes

1. Federal German Parliament, Investigation committee on the future of the media in the business community and in society, "Sicherheit und Schutz im Netz," Federal Parliament Printed Paper. 13/11002, 21.

2. Bundesamt für Sicherheit in der Informationstechnik, http://www.bsi.bund.de

3. http://www.jtc1.org/sc37 (01.03.2005).

4. http://www2.din.de/index.php?lang=en (23.03.2005).

5. http://www2.ni.din.de/index.php?lang=en&na_id=ni (23.03.2005).

6. http://www2.ni.din.de/ni-37 (23.03.2005).

7. *See* http://www.bsi.bund.de/biometrie (02.03.2005) for further information on biometric basics and facial, fingerprint and iris recognition.

8. T. Weichert, *Biometrie—Freund oder Feind des Datenschutzes?*, Computer und Recht 1997, pp. 369 ff., 375.

9. H. Bäumler, *Biometrie datenschutzgerecht gestalten*, Datenschutz und Datensicherheit 1999, p. 128.

10. W. Ernestus, *. . . da waren´s nur noch 8!*, Recht der Datenverarbeitung 2002, pp. 22 ff., 22.

11. Bundesdatenschutzgesetz, BDSG; *available at* http://www.bfd.bund.de; also available in English.

12. M. Kiper, *Biometrische Identifikation*, Computer Fachwissen 8-9/1999, pp. 46 ff., 50.

13. *See also* Eidgenössischer Datenschutzbeauftragter, 9. Tätigkeitsbericht 2001/ 2002, 30 (2.2.5.): "eines der zentralen Probleme biometrischer Systeme, ein Merkmal zu widerrufen, damit es nicht missbraucht werden kann."

14. B. Schneier, *Biometrics: Truths and Fictions, Essay on the Uses and Abuses of Biometrics*, 1998, pp.1 ff. and passim.

15. *See also* the overview of security of biometric systems: B. Wirtz, *Biometric System Security, Part 1*, Biometric Technology Today (BTT, UK) February 2003, pp. 6 ff., Part 2, BTT March 2003, pp. 8 ff.

16. M. Köhntopp, *Technische Randbedingungen für einen datenschutzgerechten Einsatz biometrischer Verfahren*, pp.177 ff., 180, in P. Horster (Hrsg.): Sicherheitsinfrastrukturen-Grundlagen, Realisierung, Rechtliche Aspekte, Anwendungen; Vieweg Braunschweig/Wiesbaden 1999.

17A. Rossnagel, Rechtliche Gestaltung informationstechnischer Sicherungsinfrastrukturen, pp. 135 ff., 170 (4.5.2.4), in V. Hammer, Sicherungsinfrastrukturen-Gestaltungsvorschläge für Technik, Organisation und Recht, Springer Berlin/Heidelberg 1995

18. Der Hessische Datenschutzbeauftragte, 30. Tätigkeitsbericht 2001, 14.1.2; Deutsche Bank Research, Biometrie – Wirklichkeit und Übertreibung, Studie Nr. 28, 22 May 2002, pp. 4 und 9 ff.

19. http://www.commoncriteriaportal.org/ (24.02.2005).

20. http://www.commoncriteria.de/it-sicherheit_deutsch/entstehung.htm (24.02.2005).

21. *See* §§ 3 and 4 BSIG (http://www.bsi.bund.de) (BSI-establishing regulation).

22. A.J. Mansfield & J.L. Wayman, Best Practices in Testing and Reporting Performance of Biometric Devices, Version 2.01, August 2002, NPL Report CMSC 14/02, http://www.cesg.gov.uk/site/ast/biometrics/media/BestPractice.pdf, (15.12.2004).

23. http://www.teletrust.de.

24. TeleTrusT e.V., *Biometrics Guide—Manual of Selection Criteria*, Version 3.0 as of 2006, *available at* www.teletrust.de.

25. zu den datenschutzrechtlichen Problemen bei Biometrie s. u.a.: Bäumler, Biometrie datenschutzgerecht gestalten, DuD 1999, 128; Davies, Touching Big Brother: How Biometric Technology Will Fuse Flesh and Machine, passim; Probst, Biometrie aus datenschutzrechtlicher Sicht, 115 ff.; s. auch ULD, Positionspapier zum Antiterrorgesetz der Bundesregierung, 10 ff.; s. auch Woodward, Biometrics: Identifying Law and Policy Concerns, 385 ff., 391.

26. Weichert, Biometrie–Freund oder Feind des Datenschutzes? CR 1997, 369 ff, 375; Der Berliner Datenschutzbeauftragte, Jahresbericht 1998, 3.5.; Der Hamburgische Datenschutzbeauftragte, 18. Tätigkeitsbericht 2000/2001, 1.3., 15.

27. Simitis-Simitis § 1 Rz. 6 zu dieser Kehrseite zunehmender Automatisierung der Datenverarbeitung im Allgemeinen.

28. Gundermann/Köhntopp, Biometrie zwischen Bond und Big Brother, DuD 1999, 143 ff., 144.

29. Federal Trade Commission, When Bad Things Happen To Your Good Name, 2.

30. Kiper, Biometrische Identifikation, CF 8-9/1999, 46 ff., 50.

31. Kumbruck, Der "unsichere Anwender"–vom Umgang mit Signaturverfahren, DuD 1994, 20 ff., 28 im Hinblick auf die mögliche Erhöhung der Sicherheit elektronischer Signaturen durch die Verwendung biometrischer Verfahren und dem gleichzeitig bestehenden Widerspruch zu Freiheitsrechten.

32. s. dazu schon Albrecht, Relevanz biometrischer Verfahren im gesellschaftlichen Kontext, 85 ff., 86.

33. For a comprehensive overview of European Data Protection Law and Biometrics see the BIOVISION Privacy Best Practice on Biometrics (A. Albrecht) at http://www.eubiometricforum.com (26.02.2005).

34. http://europa.eu.int/en/record/mt/top.html (02.03.2005).

35. http://www.europarl.eu.int/charter/pdf/text_en.pdf (15.02.2005).

36. Report on the transposition of Directive 95/46/EC at: http://europa.eu.int/comm/internal_market/privacy/lawreport_en.htm (02.03.2005).

37. Directive 95/46/EC of the European Parliament and the Council of 24 October 1995 on the protection of individuals with regard to the processing of personal data and on the free movement of such data at http://europa.eu.int/comm/internal_market/privacy/law_en.htm (18.02.2005).

38. *Id.*

39. http://www.europa.eu.int/comm/privacy (05.03.2005).

40. Directive on Privacy and electronic communication: http://europa.eu.int/eur-lex/pri/en/oj/dat/2002/l_201/l_20120020731en00370047.pdf.

41. http://europa.eu.int/comm/internal_market/privacy/docs/wpdocs/tasks-art-29_en.pdf.

42. http://europa.eu.int/comm/internal_market/privacy/docs/wpdocs/2003/wp80_en.pdf.

43. http://europa.eu.int/comm/internal_market/privacy/docs/wpdocs/2004/wp96_en.pdf.

44. http://ec.europa.eu/jushie_home/fsj/privacy/docs/wpdocs/2005/wp112_en.pdf.

45. *See also* the approach of the German Federal Data Protection Commissioner to biometrics, e.g. at Der Bundesbeauftragte für den Datenschutz, 19. Tätigkeitsbericht 2001-2002, pp. 36-37 (4.2.), at http://www.bfd.bund.de; a comprehensive academical approach has been made in the phd-thesis of A. Albrecht, Biometrische Verfahren im Spannungsfeld von Authentizität im elektronischen Rechtsverkehr und Persönlichkeitsschutz, (Trade-offs between Authentication and Personal Rights in the Application of Biometrics to E-commerce, PhD-publication), Dissertation 2003, Nomos Baden-Baden, Frankfurter Studien zum Datenschutz, hrsg. von Prof. Dr. Dr. hc. Spiros Simitis

46. Grundgesetz der Bundesrepublik Deutschland, http://bundesrecht.juris.de/bundesrecht/gg/ (01.03.2005).

47. Note that all following translations are not official translations but by the author.

48. Decisions of the Federal Constitutional Court 65, p. 1ff., population census judgement of 15 December 1983.

49. For the general function of a biometric system see chapter above.

50. Decisions of the Federal Constitutional Court, 65, p. 1ff., p. 43.

51. Gesetz zur Bekämpfung des internationalen Terrorismus (Terrorismusbekämpfungsgesetz), vom 9. Januar 2002, Bundesgesetzblatt (BGBl.) 2002 Teil I Nr. 3, S. 361 ff.

52. For details see chapter below on border control and biometric enhanced passports in Germany.

53. CEN/ISSS Initiative on Privacy Standardization in Europe, Draft Final Report, 30 November 2001, 4.26.

54. Art. 29 Data Protection Working Party, WP 80, Working document on biometrics, 3.9.

55. *See also* G. Schabhueser, Technical Consequences of the Biometric Agenda of the German Government, Second International BSI-Symposium on Biometrics, BSI 2004 (Secumedia), pp. 19 ff., 21 f.

56. Gesetz zur Bekämpfung des internationalen Terrorismus (Terrorismusbekämpfungsgesetz), vom 9. Januar 2002, Bundesgesetzblatt (BGBl.) 2002 Teil I Nr. 3, S. 361 ff.

57. http://europa.eu.int/eur-lex/lex/LexUriServ/site/en/oj/2004/l_385/l_38520041229en00010006.pdf (23.02.2005).

58. For a comprehensive insight at risks and chances of the application of RFID-Chips see: BSI, Risiken und Chancen des Einsatzes von RFID-Chips (Risks and chances of the use of RFID-Chips), Secumedia 2004, www.bsi.bund.de (28.02.2005). See also Art. 29 Data Protection Working Party for the European level, Working document on data protection issues related to RFID technology, WP 105 as of January 19, 2005 at http://europa.eu.int/comm/internal_market/privacy/docs/wpdocs/2005/wp105_en.pdf (02.03.2005).

59. European Commission C (2005) 409

60. For those see above in chapter Interoperability by standardisation.

61. The documents of ICAO can be found at http://www.icao.int/mrtd, for ISO go to http://www.iso.org.

62. The following translations are not official.

63. *See also* D. Kuegler, *Risiko Reisepass?, Schutz der biometrischen Daten im RF-Chip*, c´t 2005, Heft 5, pp. 84 ff.

64. Bizer, Elektronische Signaturen im Rechtsverkehr, 37 ff., 39.

65. Palandt-Heinrichs § 125 Rz.1.

66. MüKo-BGB-Einsele § 125 Rz. 1.

67. Palandt-Heinrichs § 125 Rz. 1.

68. Es werden unterschiedliche Arten der Formen differenziert: die Schriftform nach § 126 BGB, die öffentliche Beglaubigung nach § 129, die in §§ 39 ff. BeurkG geregelt ist, sowie die notarielle Beurkundung nach § 128 BGB, die in §§ 6-35 BeurkG geregelt ist, vgl. hierzu auch Erman-Palm § 125 Rz. 2 und MüKo-BGB-Einsele § 125 Rz. 3-5; im Verfahrensrecht ist die Schriftform teilweise angeordnet, um den Zugang einer übermittelten Erklärung nachweisen zu können, so z.B. in § 166 ZPO (Zustellung durch Übergabe eines Schriftstücks); in bestimmten Fällen genügt das schriftliche Empfangsbekenntnis als vereinfachte Form des Zugangsnachweises, vgl. § 212a ZPO (Zustellung gegen Empfangsbekenntnis bei Zustellung an bestimmte Rechtspflegeorgane oder Behörden und Körperschaften des öffentlichen Rechts), das aber auch unterschrieben werden muss.

69. MüKo-BGB-Einsele § 125 Rz. 1.

70. Bizer, Elektronische Signaturen im Rechtsverkehr, 37 ff., 39.

71. BGHZ 113, 48 ff., 51; Palandt-Heinrichs § 126 Rz. 5.

72. Palandt-Heinrichs § 126 Rz. 9.

73. MüKo-BGB-Einsele § 126 Rz. 12.

74. BGH NJW 1959, 734 und NJW 1982, 1467 f., 1467.

75. Erman-Palm § 126 Rz. 11; MüKo-BGB-Einsele § 126 Rz. 14.

76. BGH NJW 1962, 1505 ff., 1506-1507.

77. BGH NJW 1997, 1254 f., 1255.

78. Mot. I S. 179, Mugdan I S. 451 f.
79. Langenbach/Ulrich, Elektronische Signaturen, 35.
80. Bizer, Digitale Dokumente im elektronischen Rechtsverkehr, 148 ff., 150.
81. Bizer, Das Schriftformprinzip im Rahmen rechtsverbindlicher Telekooperation, DuD 1992, 169 ff., 170; vgl. auch Langenbach/Ulrich, Elektronische Signaturen, 36 zu dem kulturell verankerten Charakteristikum der Papierwelt.
82. s. dazu allgemein Palandt-Heinrichs § 125 Rz. 2 bis 2c und § 126 Rz. 5 ff.; Erman-Palm § 125 Rz.1; Staudinger-Dilcher § 125 Rz 3 und § 126 Rz 12; MüKo-BGB-Einsele § 125 Rz. 6-9; vgl. auch Raßmann, Elektronische Unterschrift im Zahlungsverkehr, CR 1998, 36 ff., 37; Erber-Faller auf dem 4. Deutschen EDV-Gerichtstag, Arbeitskreis "elektronische Signaturen," zitiert bei Bergmann/Streitz, Digitale Signaturverfahren - technische und rechtliche Fragen, Jur-PC 1/96, 36 ff., 37; Köbl, Die Bedeutung der Form im heutigen Recht, DNotZ 1983, 207 ff., 210
83. MüKo-BGB-Einsele § 126 Rz. 9 bezeichnet dies auch als "Zuordnungs-funktion": dem Rechtsverkehr werde die Identität des Ausstellers vermittelt sowie die Echtheit des Inhalts bezeugt, um damit auch Schutz gegen Fälschungen Dritter zu bieten.
84. Bizer, Digitale Dokumente im elektronischen Rechtsverkehr, 148 ff., 150.
85. Bizer Digitale Dokumente im elektronischen Rechtsverkehr, 148 ff., 150-151.
86. vgl. dazu BNotK/AgV, Positionspapier zur geplanten Änderung des BGB, die der Ansicht sind, dass dies bei der Textform nach § 126b BGB nicht der Fall sei, da anders als ein herkömmliches Schreiben eine e-mail nicht unbedingt rechtzeitig wahrgenommen werde; dies ist etwa bei Erklärungen zur Erhöhung der Miete durch den Vermieter der Fall, vgl. §§ 2 II 1 und III 1 MHG, wonach dann, wenn der Mieter auf das Mieterhöhungsverlangen nicht reagiert, die Klage durch den Vermieter zulässig wird, und nach § 651 g II 3 BGB a.F. bei der Anzeige des Reiseveranstalters, dass er die vom Reisenden angezeigten Reisemängel nicht akzeptiere, und dadurch die Verjährungshemmung endet, vgl. jetzt § 203 BGB.
87. Bizer, Elektronisch signiertes Dokument, DuD 1993, 700.
88. BGH NJW 2001, 289 ff., 290; s. auch Jauernig-Jauernig vor § 116 Rz. 2
89. vgl. auch Langenbach/Ulrich, Elektronische Signaturen, 35 f., zu den (auch kulturellen) Unterschieden zwischen "Papierwelt" und "digitaler Welt"
90. Bizer, Digitale Dokumente im elektronischen Rechtsverkehr, 151.
91. Bizer, Beweissicherheit im elektronischen Rechtsverkehr, 141 ff., 151
92. BGHZ 92, 251 ff., 253-254.
93. Baumbach/Lauterbach/Albers/Hartmann-Hartmann § 129 Rz. 5, der darauf hinweist, dass der Ausdruck "bestimmende Schriftsätze" nicht im Gesetz verwendet wird, sondern aus den Motiven zur ZPO stammt.
94. Zöller-Greger § 129 Rz.1.
95. BGHZ 101, 134 ff., 137-138; s. auch GmS-OBG NJW 1980, 172 ff., 174; Baumbach/Lauterbach/Albers/Hartmann-Hartmann § 129 Rz. 10
96. BGH NJW 1987, 2588 ff., 2589; so auch Borges, Prozessuale Formvorschriften und der elektronische Rechtsverkehr, K & R 2001, 196 ff., 206.
97. BGHZ 101, 134 ff., 137-138; s. auch GmS-OBG NJW 1980, 172 ff., 174; Baumbach/Lauterbach/Albers/Hartmann-Hartmann § 129 Rz. 10
98. BGH CR 1994, 29 ff., 30: keine Schriftform bei Telefax mangels eigenhändiger Unterschrift.

99. OLG Düsseldorf, NJW 1995, 2177 zu Datex-J. (Btx).

100. NJW 1997, 1254 f., 1255; BVerwG NJW 1995, 2121 f., 2122; BAG NJW 1989, 1822 ff., 1823.

101. VG Frankfurt HessVRspr. 93, 71; VG Wiesbaden, NJW 1994, 537 ff. zu § 81 I 1 VwGO.

102. s. auch Kuhn, Rechtshandlungen mittels EDV und Telekommunikation, 253-54.

103. BGH NJW 1995, 49 ff., 50.

104. Zöller-Greger Vor § 284 Rz. 18.

105. BGH NJW 1991, 1052 ff., 1053.

106. Zöller-Greger Vor § 284 Rz. 17.

107. Teilweise hat der Gesetzgeber entsprechende Normen explizit ausgedrückt oder durch Verknüpfung gesetzlicher Tatbestände gesetzliche Grundregeln geschaffen. Ausdrückliche gesetzliche Beweislastregeln finden sich etwa in §§ 179 I, 282, 358, 636 II und 2336 II BGB.

108. BGH NJW 1996, 1828 f.; näher zum typischen Geschehensablauf BGH NJW 1997, 528 f., 529 (Wahrscheinlichkeit muss sehr groß sein): Typizität bedeutet nicht, dass die Verkettung bei allen Sachverhalten dieser Fallgruppe notwendig immer vorhanden ist, sie muss aber so häufig vorkommen, dass die Wahrscheinlichkeit einen solchen Fall vor sich zu haben, sehr groß ist, vgl. auch BGH VersR 1991, 460 ff., 461, und Greger, Praxis und Dogmatik des Anscheinsbeweises VersR 1980, 1091 ff.

109. BGHZ 100, 214 ff., 216; BGH NJW 1996, 1828 f.

110. Zöller-Greger Vor § 284 Rz. 29.

111. Zuerst RGZ 130, 359, 360; seit BGHZ 2, 1 ff. ständige Rechtsprechung des BGH.

112. Bizer in Bizer/Miedbrodt, Die digitale Signatur im elektronischen Rechtsverkehr, 136 ff., 145.

113. Roßnagel, Die Sicherheitsvermutung des Signaturgesetzes, NJW 1998, 3312 ff., 3317.

114. so Reinhardt, Die Umkehr der Beweislast aus verfassungsrechtlicher Sicht, NJW 1994, 93 ff., 96.

115. BVerfGE 75, 183 ff., 190 f.; BVerfGE 69, 126 ff., 139 f.

116. Maunz/Dürig-Schmidt-Aßmann GG Art, 103 I Anm. 9 (Lfg. 27 November 1988).

117. Bizer, Beweissicherheit im elektronischen Rechtsverkehr, 141 ff., 156.

118. Roßnagel, Die Sicherheitsvermutung des Signaturgesetzes, NJW 1998, 3312 ff., 3318.

119. Reinhardt, Die Umkehr der Beweislast aus verfassungsrechtlicher Sicht, NJW 1994, 93 ff., 97.

120. Bizer, Der gesetzliche Regelungsbedarf digitaler Signaturverfahren, DuD 1995, 459 ff., 461 bzgl. elektronischer Signaturverfahren.

121. vgl. 2. Kapitel § 4 III.

122. Roßnagel, Die Sicherheitsvermutung des Signaturgesetzes, NJW 1998, 3312 ff., 3313.

123. s. dazu im Zusammenhang mit der elektronischen Signatur Roßnagel, Digitale Signaturen im europäischen Rechtsverkehr, K & R 2000, 313 ff., 318

124. vgl. dazu auch die Erläuterungen zu Fehlerraten und Toleranzwert im 2. Kapitel § 2 V. und § 4 III. 1.

125. Gundermann/Probst, Brennpunkte des Datenschutzes, Rz. 19; Prins/van Kralingen, Making our Body Identify For Us, 4.4.

126. Albrecht, Biometrie und Recht, 97 ff., 108-109; Probst, Biometrie aus datenschutzrechtlicher Sicht, 115 ff., 123.

127. Bizer, Der gesetzliche Regelungsbedarf digitaler Signaturverfahren, DuD 1995, 459 ff., 462 bzgl. elektronischer Signaturverfahren.

128. Britz, Urkundenbeweisrecht und Elektrotechnologie, 247.

129. Roßnagel, Die Sicherheitsvermutung des Signaturgesetzes, NJW 1998, 3312 ff., 3318; vgl. ausführlich zur Geschichte der freien Beweiswürdigung Englisch, Elektronisch gestützte Beweisführung im Zivilprozess, 32 ff.

130. Englisch, Elektronisch gestützte Beweisführung im Zivilprozess, 44, bzgl. elektronischer Signierverfahren.

131. so auch Bizer in Bizer/Miedbrodt, Die digitale Signatur im elektronischen Rechtsverkehr, 136 ff., 146, zum Umgang mit digitalen Signaturen im Beweisverfahren; vgl. auch Roßnagel, Die Sicherheitsvermutung des Signaturgesetzes, NJW 1998, 3312 ff., 3315.

132. vgl. oben die Ausführungen zur Anwendung des Anscheinsbeweises beim EC-Karten-Missbrauch § 3 II.

133. Englisch, Elektronisch gestützte Beweisführung im Zivilprozess, 45.

134. vgl. Deutsch, Die Beweiskraft elektronischer Dokumente, Jur-PC Web-Dok. 188/2000, Abs. 30.

135. Zöller-Greger § 402 Rz. 1.

136. BGHZ 37, 389 ff., 393; BGH NJW 1974, 1710 f., 1710.

137. so auch die Vermutung bei Bizer/Hammer, Elektronisch signierte Dokumenten als Beweismittel, DuD 1993, 619 ff., 621.

138. Deutsch, Die Beweiskraft elektronischer Dokumente, Jur-PC Web-Dok. 188/2000, Abs. 30.

139. s. dazu 2. Kapitel § 5.

140. Roßnagel: Rechtliche Gestaltung informationstechnischer Sicherungsinfrastrukturen, S. 135 ff., in: Hammer, Volker (Hrsg.): Sicherungsin-frastrukturen–Gestaltungsvorschläge für Technik, Organisation und Recht, Springer Berlin/Heidelberg 1995, 170 (4.5.2.4).

141. vgl. Wolf/Horn/Lindacher-Wolf § 9 H 3, der für den Fall der EC-Karte darauf hinweist, dass insbesondere dann, wenn der Betreiber keine ausreichenden Sicherheitsvorkehrungen trifft, das Risiko des Missbrauchs nicht auf den Kunden abgewälzt werden darf.

142. Judgement of the German Federal Labour Court (Bundesarbeitsgericht), BAG decision of 27 January 2004-1 ABR 7/03.

Chapter 8

Biometrics in the Private Sector: Trends and Case Studies

*Rocky C. Tsai**

I. Introduction

While current legal and policy debates concerning biometrics are predominantly focused on the increasing deployment of biometric technologies by governmental entities, the most significant impact of biometrics may ultimately occur in the private sector. As biometrics continues to evolve from an exotic technology to a commonplace feature of commercial transactions and consumer products, the debate over the regulation of biometric data collection and use will inevitably expand to encompass the practices—and feared abuses—of the commercial realm. As constitutional scholar Lawrence Lessig has pointed out in the context of the Internet (another revolutionary technology), the most significant transformative forces shaping the contours of cyberspace have originated not so much from government, but from commerce. Similarly, emerging industry standards and evolving consumer attitudes should drive the fate of biometrics

---

* Rocky C. Tsai is an associate with Orrick, Herrington & Sutcliffe LLP, San Francisco, California.

as much as or even more than top-down policies and regulations promulgated by public lawmakers.

The very word "corporation" is derived from the Latin *corpus*, meaning "body." It is appropriate, then, that while the most valuable assets of the twenty-first century corporation are typically its disembodied intellectual property, those corporations are increasingly turning their attention to biometric—that is, body-based—methods of identification to meet their day-to-day business needs. In an era when the targets of theft are not so much goods and property as information and identity itself, corporations are placing a premium on reliable technologies of authentication—allowing them to ensure that their customers, their clients, and their employees actually are who they say they are.

II. Network and Physical Access Security

Authentication by means of information such as a personal identification number or a password is currently the most widespread method of controlling access to valuable corporate resources. In spite of this widespread use, information of this nature tends to be forgotten, or even worse, divulged, at which point it becomes useless as a barrier to hackers and other digital criminals. Corporate security audits routinely reveal that most employees do not adhere to their companies' carefully articulated password policies. Those policies are designed to steer employees away from easily recognized patterns in the creation and use of their informational keys. As a result, it is not uncommon for security auditors to be able to crack over 50 percent of a given enterprise's passwords within a single day, and often within a matter of hours.[1] Social scientists have reached similarly discouraging results. In the infamous "Liverpool Street" study, surveyors deployed by the Infosecurity Europe trade show found that approximately 70 percent of commuters passing through the London Underground station at Liverpool Street voluntarily divulged their computer login and password information to a complete stranger in exchange for a bar of chocolate.[2]

It is not surprising, then, that most security experts recommend that corporations implement so-called "multifactor" authentication systems, which permit access to restricted facilities or resources only after several keys are correctly applied by the prospective user or entrant.[3] The paradigmatic three-factor authentication model consists of keys from three independent, and very different, sources: first, something the user *knows*, such as a personal identification number (PIN) (authentication by means of information); second, something the user *has*, such as a card (authentication by means of property); and third, something the user inherently *is*, such as a fingerprint (authentication by means of biometrics).

Like authentication by means of information, authentication via property—for example, an access pass or security token—suffers from the infirmity that property, like information, can easily be lost or stolen, resulting in significant replacement and re-authentication costs to the corporation. As a consequence, corporations are increasingly implementing means of authentication that rely on the characteristics of an individual's body itself. Authentication by means of biometrics is in many ways a much more robust approach to the problem of verification than authentication via information or property. Unlike a password or smart card, one does not forget to bring one's retina to work on a Monday morning. Unlike a personal identification number or an access pass, a stranger cannot easily steal or connive his way to an employee's facial structure, voice patterns, or hand geometry.[4]

III. Biometric Payment Systems

According to the International Biometric Industry Association, the commercial biometrics market will be worth approximately $5 billion by 2010.[5] Approximately half of the current biometric technologies that have been brought to market consist of some derivative of fingerprint scanning. For example, Herndon, Virginia–based BioPay has enrolled over a million people in its biometric check authorization system, which minimizes the risks of bounced checks by coupling check-based transactions with a secure fingerprint scan.[6] San Francisco, California–based Pay By Touch markets an authentication technology that allows shoppers to complete their purchases by scanning their fingerprint at a point-of-sale reader after they link their fingerprints to their credit cards or bank accounts.[7] In January 2005, Charleston, South Carolina–based grocery chain Piggly Wiggly Carolina Corporation announced that it would deploy Pay By Touch's biometric payment scanners throughout Piggly Wiggly's 120 stores across South Carolina and Georgia.[8]

Besides liberating shoppers from the burdens of the cash/check/card transaction model, systems like Pay By Touch allow stores to automatically track commercially important statistics such as customer demographics and loyalty. Biometric payment systems also benefit businesses by reducing the substantial costs associated with check and credit card fraud and by enabling retailers to avoid credit and debit card merchant fees. Market researchers estimate that sales of biometric point-of-sale equipment will total approximately $68 million in 2005, and that such sales will increase to more than $200 million by 2008.[9]

IV. Personal Data Protection

Finger-scanning technology has proven useful not only for purposes of customer convenience, but also as an instrument of corporate control and compliance. For example, health-care organizations are increasingly relying on finger-scanning technology to comply with their obligations under the Health Insurance Portability and Accountability Act of 1996 (HIPAA). Pursuant to HIPAA, the Department of Health and Human Services has promulgated a series of regulations designed to promote the standardization of electronic patient health, administrative, and financial data, as well as to safeguard the confidentiality of individually identifiable health information.[10] Biometrics has proven an effective technological means to achieve the policy goals codified in the HIPAA regulations—violations of which carry fines of up to $250,000 and imprisonment.[11] In an effort to achieve optimal compliance with its stringent obligations under HIPAA, Sharp Healthcare, a San Diego–based health-care delivery network comprising over 2,500 physicians and serving approximately 3 million patients, instituted the largest biometric authentication initiative ever implemented in a health-care environment in 2003.[12] The centerpiece of Sharp's systematic authentication overhaul consisted of the installation of some 7,000 gatekeeper fingerprint readers throughout its various constituent hospitals and medical groups. In order to gain access to HIPAA-protected patient information, employees must now log on to Sharp's informational network using their fingerprint rather than via the traditional suite of passwords and access cards.

V. Travel

Commercial biometric applications are, of course, not restricted to fingerprint scanning. Indeed, the portfolio of biometric technologies available to corporations is expected to diversify considerably as the scope, sources, and volume of market demand continue to grow. For example, corporations seeking the highest degree of authentication precision are increasingly relying on iris-scanning technology, which is considered the most accurate biometric identifier. Virgin Atlantic Airways and British Airways recently extended invitations to 3,000 select frequent fliers to enroll in an iris recognition trial at Heathrow Airport in London.[13] This is the first large-scale passenger identification protocol relying entirely on biometrics. The iris recognition technology used in the trial allows passengers to bypass the customary passport inspection and interview by customs officials. Instead, passengers simply present themselves for a two-second verification based solely on the unique composition of the pigmented portion of their eyes.

VI. Corporate Data Protection

The reliability of iris scanning also makes it an ideal technology for safeguarding repositories of confidential or proprietary corporate information. For example, Vertical Screen, a company that provides specialized employee background checks and other corporate screening services, protects the highly sensitive personal information stored in its central data-processing facility (including criminal records, credit reports, and financial histories) with a two-factor authentication protocol in which employees must pass a pair of checkpoints in order to gain entry to restricted areas. After passing through the first checkpoint, which requires an access code, the individual enters a gated portal controlled by a mirror-assisted, audio-prompted interface that positions the subject's eye for digital image capture of his iris. Authentication of the subject—which takes less than two seconds—is completed by comparing the captured image with the individual's digitally stored iris patterns.

VII. Standards and Commercial Adaptation

Perhaps even more important than the diversification of biometric technologies is the complementary trend of standardization of biometric platforms. New technologies cannot make the critical phase transition from "emerging" to "pervasive" until industry standards arrive. For example, common file formats, agreed-upon protocols for data exchange, and, perhaps most important, specifications for a standard application programming interface (API) are necessary for widespread use. Nonstandard APIs substantially inhibit the expansion of new technologies by fostering "lock-in": the inefficient balkanization of proprietary systems whereby end users are wedded to a specific technology and its particularized family of applications. Common APIs eliminate the myopic, anti-competitive effects of lock-in by promoting interoperability, allowing developers to write a single application that is compatible with biometric systems marketed by different vendors.

The movement toward standards in the commercial biometrics market was formally institutionalized in April of 1998 with the formation of the BioAPI Consortium. In March of 1999, the corporations that founded the Consortium, including IBM, Compaq, and Novell, agreed to merge their efforts to develop a common biometric API with the parallel work that was being sponsored by the Information Technology Laboratory of the National Institute of Standards and Technology.[14] Currently, the BioAPI Consortium consists of 148 members worldwide, including Barclays Bank, Hewlett-Packard, Hyundai, Infineon Technologies (formerly Seimens), Kaiser-Permanente, and the National Security Agency, and represents a variety of interests from both the public and private sector.[15]

The Consortium's Steering Committee, elected annually, consists of seven members: Compaq, Intel Corporation, IriScan, Mytec Technologies, the National Institute of Standards and Technology, SAFLINK, and Unisys. In March of 2000, the Consortium released its initial (Version 1.0) biometric API specification. An upgraded version (Version 1.1) of both the specification and its accompanying reference implementation was released in March of 2001. Over 30 corporations have now developed novel biometric applications in full compliance with the specifications set forth in BioAPI Version 1.1.

The rapid development of an industry standard API is both a manifestation of the widespread appeal of biometrics among corporations throughout the world and a harbinger of a significant, if not explosive, expansion in the volume and diversity of biometric applications that will ultimately make their way into the private sector. As the market for biometric products and services continues to grow, however, and as biometric technologies of identification and control become entrenched in the workplace, the bank, the hospital, and even the home, concerns over the security and privacy of biometric keys will inevitably grow among consumer advocacy groups, exponents of civil liberties, and ultimately public lawmakers themselves. Given these foreseeable developments, it is important to begin that public debate now rather than later, so that the commercial biometrics market is guided at the outset not only by robust technological standards, but by equally compelling legal and ethical standards as well.

Notes

1. *See, e.g.*, Kym Gilhooly, *Biometrics: Getting Back to Business*, COMPUTERWORLD, May 9, 2005, *at* http://www.computerworld.com/printthis/2005/0,4814,101557,00. html ("'Within 30 seconds, we had identified probably 80% of people's passwords,' says Fowler, whose group immediately asked employees to create strong passwords that adhered to the security requirements. A few days later, the team ran the password cracker again: This time, they cracked 70%.").

2. *Passwords Revealed by Sweet Deal*, BBC NEWS, *at* http://news.bbc.co.uk/1/hi/technology/3639679.stm (last visited January 19, 2006).

3. *See, e.g., Authentication in an Electronic Banking Environment*, FINANCIAL INSTITUTION LETTERS, Aug. 8, 2001, *at* http://www.fdic.gov/news/news/financial/2001/fil0169a.html.

4. Although biometric sensors are not foolproof, they are very difficult to trick.

5. Beth Mattson-Teig, *Biometrics Make Security Physical*, BUS. J., Sept. 8, 2000, *at* http://www.bizjournals.com/twincities/stories/2000/09/11/focus3.html.

6. Christian Meagher, *Biometric Payment Companies Claim to Have the Touch*, June 3, 2004, *at* http://www.insideid.com/ecommerce/article.php/3362971.

7. *Id.*

8. *Piggly Wiggly Expands Pay by Touch Deal*, SAN FRANCISCO BUS. TIMES, Jan. 17, 2005, *at* http://sanfrancisco.bizjournals.com/sanfrancisco/stories/2005/01/17/daily8.html.

9. *Biometric Payment Systems Beginning to Catch On in U.S.*, TAIPEI TIMES, June 27, 2005, *available at* http://www.taipeitimes.com/News/worldbiz/archives/2005/06/27/2003260988.

10. *See* 42 U.S.C. § 1320d *et seq.*; 45 C.F.R. §§ 160.101 *et seq.*

11. 42 U.S.C. § 1320d-6(b)(3).

12. *Sharp Healthcare Goes Biometric*, HEALTH DATA MANAGEMENT, March 11, 2003, *at* http://www.healthdatamanagement.com/html/PortalStory.cfm?type= newprod&DID=9875.

13. Gary Stoller, *Dulles, JFK Test Iris-Recognition Systems*, USA TODAY, March 10, 2002, *available at* http://www.usatoday.com/tech/news/2002/03/11/iris-technology. htm.

14. BioAPI Consortium, History of the API and Relationship to Other Standards, *at* http://www.bioapi.org/history.html (last visited July 26, 2005).

15. BioAPI Consortium, Consortium Members, *at* http://www.bioapi.org/members. html (last visited July 26, 2005).

Chapter 9

The Application of Biometrics to Payment Verification

Taryn Lam*

I. Introduction

We've all been there. You go on your weekly trip to buy groceries. You push your loaded cart to the checkout lane where you watch precious minutes of your free time dwindle away as you stand in line. You try to wait patiently as the person ahead of you fumbles through her purse for her checkbook, attempts to locate a pen, and fills out her check. The store cashier takes the check and requests identification. The minutes continue to tick away as the customer again rifles through her purse for her wallet and driver's license and as the clerk runs the check through the machine and awaits the bank's approval. Finally, it is your turn. To your dismay, you realize that you have left your credit card at home and must rely on either your checkbook or debit card to pay for your groceries. After mentally running through the list of remaining errands that must be completed within the next hour, you search through your wallet for your debit card, hoping to save time and avoid the drawn-out check-processing procedure that you just witnessed. You swipe the card and struggle to re-

---

* Taryn Lam is an associate with White & Case LLP, Palo Alto, California.

member your personal identification number, succeeding only after two attempts at punching the number into the keypad. You continue to wait as your bank's computers process the debit request and transmit approval to the store's computers. At long last, you breathe a sigh of relief as your payment is processed and you can leave the store.

Thanks to advancements in biometrics, the above scenario could soon become a relic of the past, as technologically outdated and unnecessary as buying carbon paper for one's typewriter. Although biometrics have traditionally been used for security measures, retail stores are beginning to utilize biometrics technology to verify payments made by their customers. Instead of undergoing the usual frustrations, shoppers can zip through the checkout lane after completing payment in mere seconds.

Biometric technology has been around for decades, but interest in such technology sharply rose after the September 11 terrorist attacks. Fingerprint scans, iris scans, face scans, and voice recognition programs, *inter alia*, were touted as potentially powerful instruments to bring about greater national security. The U.S. government and private businesses boosted investment in these areas as they became aware of the possible uses of biometrics as a tool to identify terrorists as well as in commercial applications. The increased investment led to many advancements in biometric technology, and the costs and reliability of such technology have improved to the point that the technology has become viable for widespread use by retailers for payment verification. Fingerprint scans in particular have become the most widely utilized type of biometrics in the retail setting because of the relatively low costs.[1] As a result of the application of such technology, the checkout process at such retailers has become streamlined, leading to a faster and more convenient shopping experience for consumers. The retailers themselves are also reaping myriad benefits from such use of biometrics, including increased sales and greater cost savings. This chapter will explain the ways in which biometrics can be used to facilitate retail payments and will explore the different ways in which businesses can benefit from such use.

II. How Biometric Technology Can Be Used to Verify Payments

Numerous studies have indicated that American consumers prefer to pay for their purchases electronically. A recent study conducted by the American Bankers Association revealed that since 2003, American consumers are relying more on credit cards and debit cards than on cash or checks to pay for their purchases.[2] In 2001, 57 percent of monthly payments by consumers were in the form of cash or checks; by 2005, that figure had fallen to 45 percent.[3] The next advancements in the electronic payment evolution appear to be biometrics-based payments and contactless smart cards (i.e., cards embedded with microchips that allow the user to complete a transaction by waving the card in front of an electronic card reader). In 2005, almost $3.2 billion in transactions involved these two new forms of technology.[4]

Until recently, the two leading companies that offer biometric technology in conjunction with payment verification were BioPay, based in Herndon, Virginia, and Pay By Touch, based in San Francisco, California. BioPay was founded in 1999 and specializes in the use of biometrics for check cashing and payments.[5] Its technology is available at 1,600 retail stores, and its customers range from pharmacies to national banks.[6] Pay By Touch began its biometric services in 2003 and offers biometric payments for many large retailers.[7] Pay By Touch's biometric payment technology is utilized in almost 300 retail stores.[8]

On December 6, 2005, Pay By Touch announced its intent to acquire BioPay for $82 million.[9] At the date of the announcement, BioPay had two million enrollees in its biometric programs and had processed more than $7 billion in transactions.[10] For its part, Pay By Touch could count as its customers two of the top five grocers in the United States and 55 of the top 100 retailers.[11] The success of these two companies indicates the potential for explosive growth in the field of biometrics-based payment systems.

A. Payment Verification

Consumers are encouraged to enroll in biometric payment programs because of the ease and speed of the sign-up process. Pay By Touch offers free enrollment in its program to consumers, who can choose to enroll in-

store or online via the Pay By Touch Web site.[12] Customers can complete
the enrollment process in only a few minutes, and the one-time enrollment
allows the customer to use the Pay By Touch service for free at any participating retailer.[13]

In order to begin the enrollment process, the shopper merely scans her
fingers at a special kiosk in a participating store and chooses a search code
(usually the shopper's home telephone number) that is used to assist the
system in locating the fingerprint.[14] The scanner takes an image of the
fingers; software converts the image into specific points of data, which, in
turn, are processed using a mathematical algorithm.[15]

Pay By Touch's software converts the fingerprint image into 35 to 40
points of data, which represent 10 percent of the finger.[16] The points, rather
than the fingerprints themselves, are then stored in computer systems with
"military-level encryption."[17] The stored information is secure from identity thieves because such points of data cannot be reverse-engineered back
into the shopper's fingerprint.[18] In contrast, BioPay does store images of
the customer's fingerprints in its database, in order to avoid the need to re-
enroll customers in the event of technological changes.[19] However, a BioPay
representative argues that, unlike a Social Security number, there is no
danger that a thief would be able to profitably use such fingerprint images.[20] BioPay also argues that its system is secure because it does not
allow store clerks access to customers' sensitive information and because
the customer does not leave behind any personal information with the
merchant.[21] It is likely that any security risks associated with the BioPay
system will be rendered moot by Pay By Touch's acquisition of BioPay,
because the BioPay system may be converted to the more secure Pay By
Touch system.

Along with the fingerprint scan, the shopper also provides her personal information from a government-issued identification card (such as a
driver's license) and provides the financial information that she wishes to
associate with her fingerprint by scanning her debit card(s) and/or credit
card(s), and any store loyalty or membership cards.[22] The shopper can also
scan one of her checks to link her fingerprint with her checking account.[23]

When the shopper later makes a purchase, she swipes her finger on an
optical scanner located next to the usual payment terminal and enters her
search code.[24] The system compares the scanned print against the stored

data points. If there is a match, the payment options stored during the enrollment process are displayed on the terminal screen and the shopper chooses
her preferred method of payment by pressing the corresponding button on
the payment terminal.[25] Thus, the entirety of the biometric payment process
consists of a fingerprint scan and some pushes of a few buttons. The biometrics user need not sign any sales slip, remember any complicated PIN,[26]
write any checks, present any identification card, or swipe any credit card.

B. Check Cashing

Biometrics can also be used in the retail setting to facilitate the cashing
of paychecks. BioPay's most popular product is called Paycheck Secure,
which is aimed at check-cashing stores and "non-banking" consumers.[27]
To enroll, the customer provides identification information to a teller at the
store, who enters the information via a terminal into the BioPay system.[28]
The teller then takes a picture of the consumer and scans her fingerprint.[29]
Similar to the Pay By Touch enrollment process, the BioPay Paycheck
Secure customer then chooses and enters a search code that would assist
the system in locating her fingerprint.[30] When the customer later wishes to
cash a paycheck, she simply scans her finger while the cashier scans her
check.[31] The same procedure applies if the customer wishes to pay for her
purchases by check. In both instances, the Paycheck Secure system processes the finger scan and check scan and verifies the authenticity of the
check or the amount of funds available in the customer's checking account.[32] The system terminates the transaction if the customer attempts to
pass a "bad check" or if there are insufficient funds in the checking account.[33] With more than 1.5 million members, the Paycheck Secure system is the most widely used retail biometric system in the United States.[34]

III. The Benefits of Biometric Payment Systems

The costs of biometrics technology have greatly decreased since its
inception. In 2000, the cost of a fingerprint scanner was approximately
$2,000.[35] Today, that cost has decreased to approximately $50 per scanner.[36] The cost of installing the entire Pay By Touch system would be only
a few hundred dollars per checkout lane (assuming that the lane is already
equipped with the conventional credit card swipe terminal).[37] The accu-

racy of biometric technology has similarly undergone great improvement. Pay By Touch estimates that the accuracy of its finger-scanning payment system is around 95 percent.[38] A company representative also notes that any failed scan would only result in denial of access to the customer's account; the system would never confuse two different fingerprints.[39]

The benefits of biometric payment technology can often outweigh its costs, making the technology a cost-effective investment for the retailer. As the following sections will describe, the implementation of biometric payment systems can increase sales and decrease costs. As a result, the retailer could quickly recoup the costs of its initial investment many times over.

A. Increased Sales

There are several ways in which a retailer's implementation of a biometric payment system in its stores may lead to increased customer sales. The primary benefit of a biometric payment system is the greater customer convenience resulting from the faster and easier payment process. Such increased convenience often translates to a significant increase in sales because satisfied customers are more likely to return to the same retailer for future purchases. A biometric payment system also decreases the transaction time for each sale, which allows the retailer to process a greater amount of sales in the same period of time. Furthermore, biometric payment systems facilitate the execution of store loyalty programs, which will likely lead to increased sales.

1. Greater Customer Convenience and Satisfaction

Users of biometric payments often experience a more enjoyable shopping trip due to the increased convenience. At stores with biometric payment systems, the customer does not need to remember any complex personal identification numbers or fumble for cash, a credit or debit card, or a checkbook. The cashier does not need to check identification or request the customer's signature. As a result, customers will encounter fewer delays and less hassle when paying for their purchases.

Customers will also appreciate the greater security that comes with biometric payments. Because the system is based on fingerprints that are unique to each individual, it would be very difficult for identity thieves to fool the system and make unauthorized purchases. Further, the require-ment of a fingerprint scan would act as a deterrent to potential identity thieves, who would not want records of their fingerprints to be created and later given to law enforcement officials. Adding to the security of biometric payments, the sales clerk is not given access to any of the customer's financial information, such as the full credit card or checking account number, during the purchase process.

In addition to acting as a tool against identity theft, biometric payments lead to greater customer security against physical theft of cash, credit or debit cards, or checkbooks. Because there is no need for the biometric payment user to carry these items on her person, she does not have to worry about losing such items to thieves or muggers.

Finally, the novelty of biometric technology can transform shopping from a tedious chore to a fun and exciting experience for shoppers. Cus-tomers would enjoy using futuristic technology that they have previously only read about in newspapers and magazines or seen in popular movies such as *The Minority Report*.

Businesses that have implemented biometric payment systems have testified to the higher customer satisfaction that can be attributed to such systems. Thriftway, a grocery store chain in Washington that is one of Pay By Touch's oldest customers, found the system to be a hit with its custom-ers. After introducing the system, Thriftway observed that customers were coming into its stores to use the biometrics technology; in fact, one man claimed to have driven 400 miles in order to use the technology.[40] The biometric payment system was so successful that Thriftway decided to make the system a permanent part of its stores.[41]

Similarly, Piggly Wiggly Carolina, the South Carolina subsidiary of a large national grocery store chain, received much praise from its custom-ers after testing biometric payments at four of its stores. A store represen-tative reported: "[We] found that our guests loved paying for their grocer-ies with a quick finger-scan because of the convenience of not having to fumble with wallets at check-out. . . . We also found that frequent Pay By Touch users saw Piggly Wiggly as more dedicated to providing better cus-tomer service."[42] In fact, the customer reaction was so enthusiastic that Piggly Wiggly Carolina made the decision to install the biometric payment technology at every one of its 118 stores.[43] Albertson's, the number two supermarket chain in the nation, also received "very positive" feedback from its customers after testing the biometric payment technology at a few of its stores.[44]

2. Faster Transaction Time

The length of time that it takes a retailer to process a sale depends on the type of payment used by the customer. A transaction in which the customer pays by writing a check takes an average of 64 seconds to be completed.[45] A transaction involving a credit card takes 40 to 45 seconds to be processed; a transaction involving a debit card takes 35 seconds on average.[46] The average transaction time for a cash payment takes approximately 29 seconds.[47] In contrast, the biometric payment user can complete the payment process in only 14 seconds because the entire payment process is compressed into only a fingerprint scan and a few pushes of the buttons on the payment terminal.[48] Biometrics can also substantially shorten the transaction time where the purchase includes age-sensitive products such as alcohol or cigarettes, or where the customer claims an age-based discount (e.g., senior discount), because the system can automatically verify the customer's age.[49]

The decreased transaction times resulting from the use of a biometric payment system can make a substantial difference in a business's bottom line. Shorter transaction times allow a store to process a higher number of transactions and improve the customer's shopping experience. In addition, studies indicate that the use of biometric payments can cause a 30 percent increase in impulse buys at checkout.[50]

The above benefits are not merely theoretical but are actually being realized by the companies that have implemented biometric payment systems in their stores. For example, a video store chain that installed the Pay By Touch system thereafter experienced increased sales due to the way in which the system expedited the checkout process.[51] Since the average video store customer tends to forget her membership card, the video checkout process sometimes necessitates the clerk searching through the store's electronic records to locate the customer's account.[52] The biometric payment system eliminates this need because a single fingerprint scan allows the cashier to locate the customer's membership account and allows the customer to pay for her video purchase or rental.[53]

3. *More Effective Customer Loyalty Programs*

Biometric payment technology can also help businesses generate more repeat business from existing customers. Many retailers, especially super-markets, have customer loyalty programs that provide customers with store-specific discounts. Biometrics can increase participation in such programs and make them more effective.

Biometrics can make participation in a store's loyalty program easier because the customer's fingerprint can be linked to her loyalty program account, thus eliminating the need for the customer to carry loyalty cards. With one swipe of the finger, the customer can pay for her purchase and, at the same time, activate her loyalty program membership. The customer will appreciate the fact that the loyalty program discounts will be auto-matically applied to her purchases, and the retailer will benefit from the additional data about its customers' shopping habits and preferences. Re-tailers can use the data to send coupons for the customer's favorite prod-ucts, thereby encouraging the customer to return to the store for future purchases, or to suggest new products for her to try.

One of the earlier applications of biometric technology in the retail sec-tor involved a fast-food restaurant's customer loyalty program. Hi-Tech Burrito, a fast-food restaurant in Berkeley, California, was utilizing the stan-dard stamp cards, wherein customers were rewarded with stamps for their purchases and could later redeem the stamps for free food.[54] In 1998, Hi-Tech Burrito decided to test a new program in which customers could use finger scans to pay for their food and to enroll in the loyalty program.[55] After a year, Hi-Tech Burrito noted that between 65 percent and 75 percent of its regular customers had enrolled in the biometrics program.[56]

A more recent but similar use of biometrics has led to equally success-ful results. After installing biometric payment systems in its stores and integrating its loyalty programs into the system, Piggly Wiggly Carolina discovered that its customers were returning to its stores more often, caus-ing sales to increase.[57]

B. Cost Savings to Businesses

Although, as noted above, installing a biometric payment system such as the Pay By Touch system may cost a few hundred dollars per checkout lane, businesses are discovering that the system quickly pays for itself. There are various ways in which usage of biometric payment technology can result in cost savings to a business, including lower transaction costs and fewer incidents of fraud.

1. Lower Transaction Costs

The biggest impact that a biometric payment system would have on a retailer's costs would probably take the form of lower transaction costs. In order to accept credit card payments, a merchant must pay fees such as interchange fees, switch fees, and card association fees. Interchange fees are fees paid per transaction by the merchant to the credit card company's issuer bank.[58] Switch fees are flat fees paid by the merchant to the credit card network for use of the network infrastructure.[59] Switch fees vary by transaction volume.[60]

One main benefit of biometric payments is that such payments incur much lower interchange fees than do traditional noncash payments. For a $50 credit card transaction, the average interchange fee can be as much as 87 cents.[61] For a $50 debit card transaction, the interchange fee can be up to 68 cents for an off-line transaction (one in which the debit card is used with a signature and the transaction is processed over a credit card network) and up to 45 cents for an on-line transaction (one in which the debit card is used with a PIN and the transaction is processed over a debit card network).[62] While checks are less expensive for the merchant to process because no interchange fees are involved, merchants still must pay anywhere from 30 cents to 60 cents per transaction (this amount includes costs associated with check verification, fraud prevention services, and employee time devoted to the deposit and balance of the checks received).[63]

In stores with a biometric payment system, the customer is able to pay for her purchase with a deduction from her checking account. From the customer's point of view, the effect of such a payment is the same as if she had used a debit card. However, such a payment would be processed over the Automated Clearing House network, which charges transaction fees that are much lower than those charged by credit card companies or banks.[64] For example, a Pay By Touch transaction would incur an interchange fee of 12 to 14 cents; a BioPay transaction of $100 would incur a fee of approximately 15 cents.[65] Thus, the transaction fees for biometric-based debit deductions can be up to 85 percent lower than the fees for credit card payments.[66] In addition, the merchant can encourage customers to use such a lower-cost payment method because the merchant is able to choose the methods of payment that it wishes to make available through the biometrics system and can dictate the order in which the different payment options are presented to the customer on the terminal screen.[67] In this way, one of Pay By Touch's customers, Thriftway supermarkets, was able to successfully shift its customers toward this method of payment instead of the higher-cost credit and on-line debit card methods.[68] The resulting cost savings can make a big difference for businesses with a high volume of transactions (such as supermarkets) and thin profit margins.

2. Fewer Instances of Fraud

A biometric payment system can also increase a merchant's profits because the system can lower the costs associated with preventing and correcting fraudulent transactions. Biometric payments are more secure than conventional forms of payment because the payer's identity is verified through a fingerprint rather than through a signature, which can be forged, through a PIN, which can be stolen, or through possession of a credit or debit card, which can also be stolen. A recent survey indicated that 51 percent of Americans believed that finger scans were a more secure identification method than passports, credit cards, picture identification cards, birth certificates or signatures.[69]

The system would also help stores to avoid check fraud, including "kiting," in which the thief makes multiple purchases in an area within a short period of time with the bad checks and then departs the area before the checks can bounce.[70] Under the biometric payment system, because authorization for checks occurs almost instantaneously, the system would detect the very first bounced check and would not allow any additional checks to be processed.[71] As a result, the merchant would not need to be as concerned about check fraud.

The businesses that have installed biometric systems appear to be satisfied with the effects on fraud. In addition to lower transaction costs as noted above, Thriftway experienced a substantial drop in fraudulent transactions after it installed biometric payment systems in its stores.[72] Another store owner described his experience with BioPay's Paycheck Secure:

"Because we're a neighborhood grocer, there are a lot of locals who come here to cash their checks. Prior to setting up the system, we had about 15 to 20 checks per month that were fraudulent. Now, we're down to zero."[73]

IV. Current Implementation of Biometric Payments in the Retail Sector

Although the application of biometrics to retail payments is a relatively new phenomenon, there are already many retailers who have tested such technology and experienced the attendant benefits.

As described above, groceries stores such as Piggly Wiggly Carolina and behemoth Albertson's have installed the technology and received positive responses from their customers, who have shown a strong willingness to give biometrics a try. In fact, mere months after it introduced biometric payments into its stores, Piggly Wiggly Carolina estimated that between 15 and 20 percent of customers who usually utilized a noncash method to pay for purchases had signed up for biometric payments.[74] Senior citizens, in particular, instead of shying away from the new technology, flocked to the biometric systems due to the resulting freedom to shop without carrying cards or cash that could be lost or stolen.[75] After Piggly Wiggly Carolina finished installing the biometric systems in all of its stores, it encountered even higher enrollment rates, with approximately 30 to 40 percent of its customers enrolling in the systems.[76] The company also found that the enrolled customers were shopping more often at the stores and spending more money on each shopping trip.[77]

Other retailers across the country are also implementing biometric payment systems in their stores. Lowe's Food Stores, Inc. decided to offer biometric payment and check cashing to its customers in early 2005.[78] Cub Food West Region decided to add the systems to all of its 65 stores in Minnesota after the positive customer response to its pilot program.[79] Bigg's markets in Ohio recently announced that it was launching the system in two of its stores.[80] In late 2005, Farm Fresh Supermarkets finished the installation of the systems in all of its 41 stores in North Carolina.[81] Retailers discovered that their customers were very receptive to the new technology after they received explanation of how the technology worked and its possible benefits.[82]

To keep up with the anticipated increase in demand for biometric payments, Pay By Touch has plans for dramatic expansion in the near future. According to its chief operating officer, the company expects at least 6,000 retail stores to be using Pay By Touch's biometric payment technology by the end of 2006.[83] Moreover, in October 2005, Pay By Touch announced that its investors have pledged $130 million toward the company's expansion plans.[84]

Despite its successes, biometric payment technology may encounter some obstacles to widespread use. First, not everyone can use such technology. A small percentage of the American population has naturally unreadable fingerprints.[85] Other people, such as construction workers, have very calloused hands that also cannot be read by fingerprint scanners.[86] There are also privacy concerns surrounding the theft of fingerprint data. Although theft and use of another's fingerprint would presumably be more difficult than theft and use of another's name, Social Security number, or other identifying information, the repercussions of fingerprint theft would be much more serious and more difficult to correct. After all, unlike a credit card account or a Social Security number, a person's fingerprint cannot be replaced with new prints. Finally, contactless payment cards are threatening to become the more popular technology for retail payments.[87] Nonetheless, the potential benefits that biometric technology may confer on both consumers and retailers ensure that biometric payments will continue to draw much interest.

V. Conclusion

There is no denying the strong growth of biometric payment technology, and retailers are just beginning to realize the variety and extent of the benefits that can be derived from such technology. As noted previously, over two million Americans have enrolled in BioPay's biometric system, indicating that consumers are more than ready for the new technology. The number of biometric payment users is expected to rise even higher than current levels, and investment in the technology is also on the rise. The market for biometric payment systems is expected to increase to $440 million, or 8.4 percent of the total market for biometrics, by 2010, up from $31 million, or 2 percent of the market in 2005.[88] If realized, such growth

will drastically alter the way that consumers pay for their purchases and
the way that retailers accept payments for their goods and services.

Notes

1. Susan Reda, *Brave New World of Biometrics*, STORES MAGAZINE, May 2002, *available at* http://www.stores.org/archives/may02cover.asp.

2. M.P. Dunleavey, *In the Blink of an Eye, You've Paid*, N.Y. TIMES, Dec. 17, 2005, *available at* http://www.nytimes.com/2005/12/17/business/17instincts.html?ex=1136523600&en=fca57fee93ef2e1c&ei=5070.

3. *Id.*

4. *Id.*

5. *See* Christian Meagher, *Biometric Payment Companies Claim to Have the Touch*, June 3, 2004, http://www.insideid.com/ecommerce/article.php/11782_3362971_1.

6. *Id.;* Jonathan Birchall, *Pay By Touch System Set to Expand in 2006*, FINANCIAL TIMES, Dec. 26, 2005, *available at* http://news.ft.com/cms/s/97146656-762c-11da-a8a9-0000779e2340.html.

7. Meagher, *supra* note 5.

8. Birchall, *supra* note 6.

9. Press Release, Pay By Touch, Pay By Touch to Acquire BioPay, Strengthening Both Companies' Value to Retailers and Consumers (Dec. 6, 2005), http://www.paybytouch.com/news/pr_12-06-05.html.

10. *Id.*

11. *Id.*

12. Pay By Touch, http://www.paybytouch.com/whatis/enroll.html (last visited Jan. 17, 2006).

13. *Id.*

14. Meagher, *supra* note 5.

15. *See Pay By Touch Helps the World Go Walletless With Finger-Scanning Payment Systems*, PR NEWSWIRE, July 22, 2004, http://www.prnewswire.com/cgi-bin/stories.pl?ACCT=104&STORY=/www/story/07-22-2004/0002215824&EDATE=.

16. *See* Ryan Holeywell, *Retailers Testing Biometric Payments*, WASH. TIMES, June 10, 2005, *available at* http://www.washtimes.com/upi-breaking/20050610-021309-8168r.htm.

17. Kathy Chu, *Will That Be Cash, Credit – or Finger?*, USA TODAY, Dec. 1, 2005, *available at* http://www.usatoday.com/tech/news/techinnovations/2005-12-01-cash-credit-finger_x.htm.

18. *See* PR News, *supra* note 15.

19. Chu, *supra* note 17.

20. *See* Holeywell, *supra* note 16.

21. *See id.; see also* Robert Lemos, *Fingerprint Payments Taking Off Despite Security Concerns*, THE REGISTER, Oct. 8, 2005, *available at* http://www.theregister.co.uk/2005/10/08/fingerprint_payments.

22. Pay By Touch, http://www.paybytouch.com/merchants/wallet.html (last visited Jan. 17, 2006).

23. *Id.*

24. *See* Michele Chandler, *Point of Sale: Retailers Try Their Hand at Finger-Scanning Payment System*, MERCURY NEWS, June 20, 2005, *available at* http://www.mercurynews.com/mld/mercurynews/business/11938666.htm?template=contentModules/printstory.jsp

25. *See id.*

26. The search code is usually the shopper's home telephone number, which is easier to remember than the random string of numerals that form most PINs.

27. Meagher, *supra* note 5.

28. *Id.*

29. *Id.*

30. *Id.*

31. *Id.*

32. *Id.*

33. *Id.*

34. *Lowe's Foods Brings Biometric Fingerprint Payments and Check Cashing to Customers,* LINCOLN TRIBUNE, March 22, 2005, *available at* http://www.lincolntribune.com/modules/news/print.php?storyid=978.

35. Holeywell, *supra* note 16.

36. *Id.*

37. Chandler, *supra* note 24.

38. Meagher, *supra* note 5.

39. *Id.*

40. Jo Best, *Shoppers Can Pay for Their Groceries With the Touch of a Finger*, SAN FRANCISCO CHRONICLE, Feb. 2, 2005, *available at* http://sfgate.com/cgi-bin/article.cgi?file=/chronicle/archive/2005/02/02/BUG7QB413P1.DTL&type=printable.

41. *Id.*

42. Payment News, *Pay By Touch to Deploy in All South Carolina, Georgia Piggly Wiggly Stores,* May 12, 2005, http://www.paymentsnews.com/2005/05/pay_by_touch_to.html.

43. *See* Andrea Orr, Piggly Wiggly Finds the Right Touch, July 18, 2005, http://www.extremenano.com/article/Piggly+Wiggly+Finds+the+Right+Touch/155722_1.aspx

44. Grace Wong, *Cash or Plastic? How About Fingerprint?*, MONEY, July 19, 2005, http://money.cnn.com/2005/07/19/pf/security_biometrics/.

45. Arik Hesseldahl, *One-Fingered Discount at the Grocery Store*, FORBES, June 17, 2005, *available at* http://www.forbes.com/2005/06/17/digital-life-fingerprint-scanners-cx_ah_0617diglife_print.html.

46. *Id.*

47. *Id.*

48. *Id.*

49. *See* Pay By Touch, http://www.paybytouch.com/merchants/age_verification.html (last visited Jan. 17, 2006).

50. LoveToKnow Business, http://business.lovetoknow.com/wiki/Credit_Card_Processing:_Biometrics (last visited Jan. 17, 2006).

51. *See* Meagher, *supra* note 5.

52. *Id.*

53. *See id.*

54. Geneva Rinehart, *Biometric Payment: The New Age of Currency*, March 2000, http://www.hotel-online.com/News/PressReleases2000_1st/Mar00_Biometric Currency.html.

55. *Id.*

56. *Id.*

57. Orr, *supra* note 43.

58. *Payment Types at the Point of Sale: Merchant Considerations*, PAYMENTS SYSTEM RESEARCH BRIEFING (Federal Reserve Bank of Kansas City), Dec. 2004, at 2, http://www.kc.frb.org/FRFS/PSR/PSR-BriefingDec04.pdf.

59. *Id.*

60. *Id.*

61. *Id.*

62. *Id.*

63. *Id.* at 1.

64. Holeywell, *supra* note 16; *see also* Chandler, *supra* note 24.

65. *See* Hesseldahl, *supra* note 45; Payment News, *More Grocers Sign With BioPay*, Nov. 21, 2005, http://www.paymentsnews.com/2005/11/more_grocers_si.html#more.

66. Holeywell, *supra* note 16.

67. Pay By Touch, http://www.paybytouch.com/merchants/payment.html (last visited Jan. 17, 2006).

68. Digital Transactions, *Pay By Touch Close to Breakthrough Deals, CEO Says*, Feb. 11, 2004, http://www.digitaltransactions.net/newsstory.cfm?newsid=156.

69. Payment News, *supra* note 65.

70. Meagher, *supra* note 5.

71. *Id.*

72. Best, *supra* note 40.

73. Reda, *supra* note 1.

74. Orr, *supra* note 43.

75. *Id.*

76. Chu, *supra* note 17.

77. *Id.*

78. *Supra* note 34.

79. Secure ID News, http://www.secureidnews.com/news/2005/10/17/cub-foods-west-region-launches-biometric-payment-technology-in-all-stores/ (last visited Jan. 17, 2006).

80. Press Release, Pay By Touch, Bigg's Launches Pay By Touch Technology (Nov, 4, 2005), http://www.paybytouch.com/news/pr_11-04-05.html.

81. Press Release, Pay By Touch, Farm Fresh Launches Pay By Touch Biometric Payment Technology in All Stores (Dec. 16, 2005), http://www.paybytouch.com/news/pr_12-16-05.html.

82. *See* Reda, *supra* note 1.

83. Birchall, *supra* note 6.

84. Lemos, *supra* note 21.

85. Wong, *supra* note 44.

86. Orr, *supra* note 43.

87. *See* Chu, *supra* note 17

88. Lemos, *supra* note 21.

Chapter 10

Biometrics and Digital Rights Management

Chris Jay Hoofnagle*

I. Biometrics Overview

Biometric identification systems are automated methods of recognizing a person based on one or more physical characteristics, such as fingerprints, voice, or facial characteristics. Computer-based pattern matching is at the core of all biometric systems. The technologies available are subject to varying degrees of error, which means that there is an element of uncertainty in any match.

The accuracy of biometric systems is measured by their false acceptance and false rejection rates. A false acceptance is when the wrong individual is matched to a stored biometric. A false rejection is when an individual is not recognized who should have been. The two measures are dependent. In reducing false acceptances, the false rejection rate will increase. Reducing false rejections will cause the false acceptance rate to go up. Most biometric systems adjust false acceptances or false rejections to the type of application and the amount of security required. High-security areas, such as bank vaults and military installations, are protected by biometric systems that minimize fraudulent acceptances. The false acceptance

* Chris Jay Hoofnagle is director of the West Coast office of the Electronic Privacy Information Center, San Francisco, California.

rate must be low enough to prevent imposters, but as a result, people who rightfully should be accepted are sometimes refused. In these cases, human intervention is typically available to provide authentication when the biometric system fails.

Fraud occurs when either an imposter is trying to be accepted as someone else to gain entry or usurp funds, or when an imposter is trying to avoid being recognized as someone already enrolled in the system and tries to enroll multiple times. The first is a form of identity theft; the second creates multiple identities for a single individual. Both types of fraud must be safeguarded against in any biometric system; however, depending on the application, it may be reasonable to relax one criterion to prevent the other.

There is no perfect biometric system. Each type of biometric system has its own advantages and disadvantages, and must be evaluated according to the application for which it is to be used.

A. Creating and Using an Identity Database

There is a distinction between authentication, identification and enrollment. Authentication is the easiest task for a biometric system to perform. Identification is more difficult and much more time-consuming. The enrollment process determines the ultimate accuracy of the biometric system. A single biometric system can be created for identification or authentication but not both, although the two applications can share the same database of biometric samples.

B. One-to-One Matching

Authentication answers the question, am I who I say I am? A person presents a biometric sample and some additional identifying data, such as a photograph or password, which is then compared to the stored sample for that person. If the person is not an imposter, the two samples should match. This is known as a one-to-one match. If a nonmatch occurs, some systems retake up to three samples from the person to find a best match. This is the simplest task of a biometric system because the independent identifiers help to corroborate the individual. The biometric acts as a secondary password to protect the individual. Authentication of an individual takes at most a few seconds.

C. One-to-Many Matching

Identification means to answer the question, who am I? A person provides a sample biometric, sometimes without his knowledge, and the system must compare that sample to every stored record to attempt to return a match. This is known as a one-to-many match, and is done without any corroborating data. Because the matching process is based on the closeness of the new sample to a stored sample, most systems return a likely list of matches. Others return a single match if the sample is similar enough. The time for the result depends on the size of the database. The FBI's Integrated Automated Fingerprint Identification System (IAFIS), which is used to identify criminals, can perform over 100,000 comparisons per second, usually completing an identification in 15 minutes with a database of over 42 million records.[1] If identification must be done on a wide scale. the number of comparisons that would need to be done simultaneously will be astronomical. In addition, consumers might be unwilling to wait more than a few seconds to be able to use their bank ATMs or on-line service.

Negative identification is when an individual can be accepted to receive a benefit only if he is not yet enrolled in a database, such as a government-run welfare program or drivers' registry. Even negative identification is susceptible to fraud. A person already enrolled in the system can avoid being recognized by attempting to falsify his biometric or skew the data collection. Rejecting imperfect images in the enrollment process improves the integrity of the database but cannot solve all enrollment problems.

D. Entering a New Person into the Database

Enrollment is the process of introducing a new person into the database. The person's biometric must be sampled and stored together with his or her identity. The greatest problem is there is no existing guarantee as to that identity. A biometric system can only be as good as the accuracy of any background information that is relied on. If fraudulent information is used to enroll an individual, through a fake birth certificate or stolen Social Security number, a biometric can only verify that the person is who they said they were at the time of enrollment. One important enrollment test is to match every new person against all other entries to check for duplicate entries and possible fraud. Without this check, once a person is in the database, it will be impossible to trace an imposter assuming multiple identities.

II. Fraud Will Not Be Eliminated Through Use of Biometrics

Because of the numerous practical, logical, and technological flaws inherent in any biometric implementation, use of biometric technologies will not serve to effectively prevent identity theft. Instead, it will create new liabilities while draining away resources and threatening the privacy of those the technology ostensibly protects.

A. There Are Too Many Practical Problems with Biometrics

Once a biometric identifier is compromised, there will be severe consequences for the individual. It is possible to replace a credit card number or a Social Security number, but how does one replace a fingerprint, voiceprint, or retina? These questions need to be considered in the design and deployment of any system of biometric identification for a large public user base.

Because biometric systems are being sold as a more effective method for authentication and verification, it follows that users of these systems will have an increased trust in, and thus reliance on, the systems' performance. As such, if a biometric is compromised, the actions of the imposter who successfully circumvented the system would be more trusted than actions by non-biometric imposters as a result of the users' perceived trust in the biometric system.

As a result, the trust placed in the effectiveness of biometric systems will act as a double-edged sword that will greatly increase the damage done to victims of identity theft. Such trust in a biometric system is harmful and counterproductive for two reasons. First, because a biometric cannot be replaced if corrupted (i.e., stolen or used for identity theft), victims will effectively be expelled from the "trusted" system, prohibiting their participation in whatever the biometric system protects. Second, this trust will likely instill a false sense of security that will prompt users to entrust more valuable information to the system. This, in turn, creates greater damage for the victim. In short, the more you entrust to the system, the more you will lose when it is corrupted.

B. It Is Too Cost-Prohibitive to Use Biometrics on a Wide Scale

A user wishing to implement a biometrics system must pay for not only the "readers," but also for the setup, installation, and maintenance of a system that must be continuously updated due to the fluid nature of biometric identifiers. Additionally, a user must be financially prepared to deal with the cost of correcting the system if it is compromised with corrupted information (e.g., a knowingly forged fingerprint or vocal reproduction).

C. Some People Will Be Unable to Enroll in Biometric Systems

There will always be a small but substantial percentage of users who cannot enroll in biometric systems either because they are unable to produce the necessary biometric (a missing finger or eye) or they are unable to provide a quality sample at enrollment. Others repeatedly cannot match their biometric to the stored template. These individuals will never be identified by the biometric system. Even if only 1 percent of the general population, which is approximately 3 million people, could not participate in a biometric system, this number is significant enough to raise serious concerns about the effectiveness of the system.

This fact not only raises the cost for biometric users by requiring additional verification tools, but also increases the liability for identity theft. By accommodating those unable to participate in the biometric system, users are unwillingly opening a backdoor to those wishing to circumvent the biometric system. Such a liability could potentially render the existence of a biometric system irrelevant.

Since some biometrics deteriorate with age, the elderly will be particularly affected. They will constitute the largest portion of those unable to enroll in or be recognized by a biometric system. There needs to be an alternative solution for those who cannot be recognized by a biometric system so that they will not be denied their rightful benefits.

D. Collection of Information Creates New Threats to Privacy

It is important to recognize in the design of any system of biometric identification that the creation of a database linked to the individual and containing access to sensitive, personally identifiable information will create a new series of privacy issues. Administrators of these systems as well as those who gain access to these databases unlawfully will have access to personal information as if they were themselves the individual subject. It is conceivable that data could be altered either by administrators or by those who gain unlawful access to the database. The result would be records that wrongly indicate biometric authentication when in fact the subject did not engage in the event recorded. There are techniques to minimize these risks, but no system is foolproof.

E. There Are Many Technical Problems with Biometrics

Biometric technology is too technologically flawed to effectively combat identity theft. In addition to inherent problems with any biometric system, the different types of biometric systems all have unique flaws, each of which are susceptible to some form of circumvention.

1. Uniqueness of Biometric Data Is Affected by Time, Variability, and Data Collection

The key to any biometric system is that the biometric being measured is unique to individuals and unchanging over time. Otherwise, the stored biometric associated with an individual needs to be periodically updated. There are several factors affecting the accuracy of any identification. Biometric data collection can be affected by changes in the environment, such as positioning, lighting, shadows, and background noise. But the biometrics of an individual are also susceptible to change through aging, injury, and disease. Because of this, the accuracy of all biometric systems diminishes over time.

2. Collecting Biometric Data Introduces Errors in the Data

Any biometric sample, whether a fingerprint, voice recording, or iris scan, is not matched from the raw data. There is too much data to store and compare during each attempted match, especially if the sample must be transmitted to a central database for matching. Instead, biometric systems use templates that represent key elements of the raw data. Face recognition systems need the most number of features to be extracted and hand scans need the least. The extracted features are compressed further into a sample template, which is then compared to a stored template to determine if there is a match. Information is lost with each level of compression, making it impossible to reconstruct the original scan from the extracted points. Since even minor changes in the way a sample is collected can create a different template for a single individual, matches are based on probability. Systems are adjustable to the amount of difference they will tolerate to confirm a match. The more independent the data available for matching, the more credible the match.[2]

3. Increasing the Speed of Biometric Systems Can Introduce Error

In extremely large populations, storage of templates is partitioned into characteristics, or bins, for ease of searching. These bins can be based on external characteristics, such as gender or race, or they can be based on the biometric's internal characteristics. Traditional fingerprint identification has been based on the binning idea, with classifications based on whorls, loops and arches. Computerized systems take advantage of this concept. While binning can speed the time for identification and allows for better statistical matches within each bin, if a template is wrongly binned, it can never be found.[3]

III. Systems Are Subject to Circumvention

There are several ways to try to circumvent a biometric system. False identification at enrollment, physically altering a personal biometric, skewing the sample collection by not cooperating, and hacking into or falsifying the database are all ways that biometric recognition can be compromised. Sample data could even be altered or stolen during transmission to a central database. How a biometric system is set up, protected, and maintained will determine the effectiveness of the system.

One often-asked question is whether biometrics can be defeated by prosthetic devices. The best biometric scanners would detect a pulse or heat from the individual to make sure that the sample has come from a live human being. However, it should be noted that if biometric systems are going to be implemented on a grand scale, it is unlikely that the "best" (i.e., more expensive) scanners will be purchased, but rather will more than likely be chosen with an eye toward budget.[4]

Additionally, a group of Japanese scientists have conducted a study whereby they were able to deceive fingerprint scanners with an astonishing success rate by using a mold made from a material similar to that which makes up "gummy bears."[5] The experiment, which tested 11 different types of fingerprint systems, found that all of the fingerprint systems accepted the gummy finder in their verification procedure more than 67 percent of the time.

IV. Digital Rights Management and Privacy

Today, individuals are free to explore different ideas presented in books, music, and movies anonymously.[6] Brick-and-mortar transactions allow individuals to purchase media with cash without leaving any personally identifiable record.[7] Similarly, many libraries have developed circulation systems that retain no transaction record once the borrowed media are returned.[8] In these systems, tracking or reporting on individuals' media consumption habits is difficult. The process is laborious and gives staff members or owners enough time to question whether the transaction information should be conveyed to law enforcement or marketers.

With the advent of easy to use peer-to-peer software applications, many Internet users have engaged in outright piracy by freely trading digital content. In a reaction to this, content owners have proposed Digital Rights Management (DRM) systems. DRMs restrict the use of digital files in order to protect the interests of copyright holders. DRM technologies can control file access (number of views, length of views), altering, sharing, copying, printing, and saving. These technologies may be contained within the operating system, within the program software, in the actual hardware of a device, or a combination of all three. It is assumed that through deploying DRM and the architecture needed to support the technology, piracy will be curbed.

DRM systems take two approaches to securing content. The first is "containment," an approach where the content is encrypted in a shell so that it can only be accessed by authorized users.[9] The second is "marking," the practice of placing a watermark, flag, or XrML tag on content as a signal to a device that the media is copy-protected.[10] Some systems combine the two approaches. Nevertheless, according to Princeton University Computer Science Professor Ed Felten, DRMs are vulnerable to cracking by individuals with moderate programming skills.[11]

Some existing DRM systems implicate privacy because they allow copyright owners to monitor private consumption of content. In an attempt to secure content, many DRM systems require the user to identify and authenticate a right of access to the protected media. In the case of Microsoft's eBook Reader, this means that the media software and users' choices in electronic books are digitally linked not only to the user's computer, but also to the company's identity management system, Microsoft Passport.[12] This arrangement allows tracking of both the individual and the individual's computer. Some systems, such as Microsoft's Windows Media Player, assign a Globally Unique Identifier (GUID) to the media device that facilitates tracking.[13] These systems create records that enable profiling and target marketing of individuals' tastes by the private sector.

A recent lawsuit illustrates how DRM implementations can be invasive to personal privacy. In February 2002, SunnComm, Inc., a DRM systems developer, and Music City Records settled a lawsuit by a California woman who objected to their practice of tracking and disclosing personal information—including music consumption patterns—to third parties with no opt-out scheme. In the case, the plaintiff's attorney, Ira Rothken, filed suit under a broad California consumer protection statute, arguing that SunnComm "never disclose[d] on the shrink-wrap of the CD(s) that consumers cannot listen to music on their computers anonymously. If left unchecked, this will be the start of an era where consumers will be coerced to give up their privacy to listen to music on their computers."[14] The settlement agreement required the companies to provide notice to consumers of their information collection practices and to refrain from requiring consumers to disclose their personal information as a condition of downloading, playing, or listening to a CD.[15]

While this settlement agreement is important, not all Americans can avail themselves of California's consumer protection laws. Given adequate notice to consumers, it is likely that other states and the FTC will not object to Sunncomm-style DRMs, and assume that the user consciously and freely accepted the invasion of his or her privacy when buying the product. Users of these new systems will be taken from a culture where there is freedom to enjoy media anonymously to one where access will be conditioned upon revealing one's identity. And once the individual has given up his or her freedom of anonymity, media companies will claim that they have the freedom to exploit information about the individual's media consumption by selling it to others—perhaps even the government.[16]

V. U.S. Law and Tradition Have Protected Privacy of Media Consumers

Traditionally, federal and state law has set out standards to protect individuals' choices in consumption of media.[17] It is understood that in this context, privacy protection provides the developmental space needed for individuals to hone skills necessary to exercise First Amendment freedoms.

Recognizing a need for insulation from outside scrutiny and interference, librarians incorporated formal policies for protecting patron privacy in 1939.[18] Also, the states have erected a framework of protections for library circulation records.[19] Congress acted to protect television viewing habits by enacting the Cable Communications Policy Act of 1984.[20] Under the act, cable operators must obtain opt-in consent before transferring user data to third parties.[21] They must also regularly destroy users' data.[22] Similarly, the Video Privacy Protection Act of 1998 was passed by Congress to create opt-in protections for those who rent media on videotape.[23] That act specifies procedures for law enforcement access to customer records[24] and requires regular data destruction.[25] Both laws allow individuals to access their data,[26] and both carry civil remedies for violation.[27]

Notes

1. *What Could Biometrics Have Done?*, http://www.biometricgroup.com/e/ Brief.htm (last visited Jul. 15, 2002).

2. James L. Wayman, *Generalized Biometric Identification System Model*, U.S. National Biometric Test Center, Proc. 31st IEEE Asilomar Conf. Signals, Systems and Computing (1997), http://www.engr.sjsu.edu/biometrics/nbtccw.pdf

3. James L. Wayman, *Large-Scale Civilian Biometric Systems*, U.S. National Biometric Test Center, Proc. CardTech/SecurTech Government (1997), http:// www.engr.sjsu.edu/biometrics/nbtccw.pdf.

4. *See, e.g.,* Lisa Thalheim, Jan Krissler & Peter-Michael Ziegler, *Body Check,* Heise Online, Nov. 2002, *available at* http://www.heise.de/ct/english/02/11/114/.

5. Tsutomu Matsumoto, et. al., *Impact of Artificial "Gummy" Fingers on Finger-print Systems,* Prepared for Proceedings of SPIE vol. #4677, OPTICAL SECURITY AND COUN-TERFEIT DETERRENCE TECHNIQUES IV, January 2002, *available at* http://cryptome.org/gummy.htm. The paper concludes that "gummy fingers, namely artificial fingers that are easily made of cheap and readily available gelatin, were accepted at extremely high rates by particular fingerprint devices with optical or capacitive sensors." *Id.*

6. *See* Julie Cohen, *A Right to Read Anonymously: A Closer Look at "Copyright Management in Cyberspace,* 28 CONN. L. REV. 981 (1996).

7. Even where a customer record is created, many booksellers will resist law en-forcement access to customers' personal information. *See, e.g., Court Overturns Book-store Ruling,* WIRED MAGAZINE (Apr. 9, 2002), *available at* http://www.wired.com/news/privacy/0,1848,51667,00.html; Daniel J. Solove, *Digital Dossiers and the Dissipa-tion of Fourth Amendment Privacy,* 75 S. CAL. L. REV. 1083 (2002).

8. *See generally* Mary Minow, *Library Patron Internet Records and Freedom of Information Laws,* 9 CAL. LIBRARIES 8 (Apr. 4, 1999).

9. Professor Edward Felten, Address at the Boalt Hall Copyright Workshop (Mar. 22, 2002).

10. *Id.*

11. *Id.*

12. This service is now called ".Net Passport." Russel Kay, *Copy Protection: Just Say No,* COMPUTERWORLD, Sept. 4, 2000; Chris Jay Hoofnagle, *Overview of Consumer Privacy 2002,* 701 PRACTICING L. INST. 1339 (2002), http://www.epic.org/epic/staff/hoofnagle/plidraf12002.pdf; Chris Jay Hoofnagle, *Digital Rights Management and Privacy,* Presentation to the Santa Clara University Law School Symposium on Infor-mation Insecurity, Feb. 8, 2002, http://www.epic.org/epic/staff/hoofnagle/drm.ppt; Megan E. Gray & Will Thomas DeVries, *The Legal Fallout From Digital Rights Man-agement Technology,* 20 COMP. & INTERNET LAW. 20 (April 2003).

13. Richard Smith, *Serious Privacy Problems in Windows Media Player for Win-dows XP,* COMPUTERBYTESMAN, Feb. 20, 2002, *at* http://www.computerbytesman.corn/privacy/wmp8dvd.htm.

14. DeLise v. Fahrenheit, No. CV-014297 (Cal. Sup. Ct. Sept. 6, 2001) (Pl. Comp. at ¶ 1), *available at* http://www.techfwm.com/mccomp.pdf.

15. Press Release, SunnComm, Inc., Sunncomm and Music City Records Agree to Resolve Consumer Music Cloqueing Law Suit by Providing Better Notice and En-hancing Consumer Privacy (Feb. 22, 2002), http://www.xenoclast.org/free-sklyarov-uk/2002-February/001580.html.

16. Daniel J. Solove, *Digital Dossiers and the Dissipation of Fourth Amendment Privacy,* 75 S. CAL. L. REV. 1083 (2002).

17. *See generally* MARC ROTENBERG, THE PRIVACY LAW SOURCEBOOK: UNITED STATES LAW, INTERNATIONAL LAW, AND RECENT DEVELOPMENTS (2002).

18. Article 11 specifies: "It is the librarian's obligation to treat as confidential any private information obtained through contact with library patrons." 1939 CODE OF ETHICS FOR LIBRARIANS, AMERICAN LIBRARY ASSOCIATION (1939), *available at* http://www.ala.org/Template.cfm?Section=Historyl&Template=/ContentManagement/Content Dis-play-cfm&Conten tlD=8875.

19. ROBERT ELLIS SMITH, COMPILATION OF STATE AND FEDERAL PRIVACY LAWS 40-41 (Privacy Journal 2002).

20. Cable Communications Policy Act, 47 U.S.C. § 551 et seq. (2003).

21. 47 U.S.C. § 551(c).

22. 47 U.S.C. § 551(e).

23. Video Privacy Protection Act, 18 U.S.C. § 2710 (2003).

24. 18 U.S.C. § 2710(b)(3).

25. 18 U.S.C. § 2710(e).

26. 47 U.S.C. § 551(d); 18 U.S.C. § 2710(b)(2)(A).

27. 47 U.S.C. § 551(f); 18 U.S.C. § 2710(c).

Chapter 11

Unintended Consequences of Biometrics

Todd Inskeep and Theodore F. Claypoole *

I. Introduction

A man walks into a bar, sits down at a booth, and is promptly greeted by an unknown server carrying a drink. "Good evening, Joe, would you like the usual, or something different tonight?"

"I'll have the usual, and can I have a menu?" replies Joe, startled that a stranger seems to recognize him.

The server sets the drink down in front of Joe and asks casually, "Would you like us to open a tab on your Visa, or will you be paying another way this evening?"

Joe looks flustered, thinks for a minute and says, "Sure, let's put it on the Visa, like last time." He stares into his drink, and then takes a sip as the server walks away.

Several minutes later, Joe notices that his menu hasn't arrived. As he starts looking around to catch the server's eye, Joe notices a policeman coming into the bar. Joe tries to get up casually, turning to avoid the

---

* Todd Inskeep is an information security architect at a large U.S. bank and regularly teaches a graduate-level information security class. Theodore F. Claypoole is a member of the law firm of Womble, Carlyle, Sandridge & Rice, Charlotte, North Carolina.

policeman's eyes, but notices the policeman is already headed straight for his booth. Resignedly he sits back down.

This scenario may seem like a part of a science fiction movie, but it is perfectly plausible within the realm of today's technology, and it exposes some of the unintended consequences associated with widespread adoption of biometrics for identification, verification, and improvement of business practices. The bar, in an attempt to improve customer service, uses a camera mounted near the door and a face recognition system to recognize customers. The customer's name, drink, and payment information is automatically routed to a server station and pops up on the screen, alerting the server to Joe's presence. The server mixes Joe's regular drink and then carries it over, hoping to give Joe a great customer experience.

Meanwhile, a flag on the bar's records automatically transmits a copy of Joe's photograph to the police as he entered the bar. The police's face recognition system also recognizes Joe, and sends a car to the bar to pick up him for an infraction. With modern electronic communications, the system could even notify the bar to serve Joe slowly so that the police would have time to arrive.

More and more often, following September 11, 2001, U.S. citizens and companies are asked to provide information for law enforcement related to terrorism. Legal experts diverge over whether these reporting requirements erode our constitutional freedoms or provide a necessary tool for fighting terrorism. Herein, we will focus on the fact that more and more companies are collecting biometric information for various purposes. Once collected, biometric information becomes yet another piece of data about the customer that companies and the government can use, subpoena, discover, and lose.

This chapter focuses on such unintended consequences and the legal ramifications of those consequences to corporations, individuals, and their lawyers. Tackling the unintended consequences requires some common understandings, and this chapter addresses the groundwork by briefly reviewing several topics. First, we will examine what companies are currently measuring and what biometric information they are collecting. Second, we will address why companies (and the government) are collecting these biometrics. Next, we will review current trends in the regulation of the legal and policy space affecting the practical use of biometrics. These introductory topics lead to the primary question: What social and legal issues, whether apparent or unanticipated, are likely to arise with the wide-spread implementation of biometric technology? In other words, when companies have collected thick files of biometric information, how might the companies, or the government, use them, thus creating unintended consequences? Finally, the chapter proposes corporate practices and regulatory and legislative approaches to managing expectations all around.

II. What Are Companies Measuring?

"Biometrics" means, literally, "life measures." Research has shown that there are many measurements that appear unique to each individual. Companies buy and sell products that measure typing patterns, fingerprints, the blood vessels in the eye, the length of a person's fingers or cheekbones, the length of a person's stride, and the speed at which he walks. Many more types of measures are available, including the DNA used to solve so many television crimes on shows like *Law & Order* and *CSI*.

Given that companies have already begun collecting measurements, we expect the acceptance and use of biometrics to expand and become more widespread. This expansion means that companies may consider biometrics from two viewpoints: permissive capture of biometric measurements and nonconsensual capture of biometric measurements from a distance.

First, companies can ask for biometrics such as fingerprints, signatures, DNA, voiceprints, retinal scans, face geometry, and more from the people who their products and services. Customers can choose whether or not to offer biometrics to organizations, much like customers can choose to sign up for affinity cards, or to pay by cash or credit. The biometric indicator might be used to offer customized services, discounts, or other incentives like cash or prizes to customers. Companies might have to provide incentives in order to obtain biometric information from customers, much as grocery stores now offer product discounts in exchange for transactional information from their customers in the form of scan cards. As with grocery transactional information, many people would probably provide biometric data in exchange for relatively small incentives and with few questions.[2] However, some people will refuse to give up their biometric readings to companies for any incentive.

A company's employees have fewer choices than its customers. Allowing one's biometric indicators to be used at work may be a requirement of employment[3] or simply an assumed condition of retaining a job. Peer pressure or the opportunity for higher pay or to be part of a special project may lure employees into giving the company their biometric information. An employee's fingerprint, retinal scan or voiceprint might be required for access to secure areas of the company: the tarmac at an airport, the vault at SunTrust Bank in Atlanta holding the secret formula for Coca-Cola, or the gold vault at Fort Knox. Or biometrics might be required for every employee to simply get through the gates surrounding the office.

Having collected the biometric measurements for one purpose, the company may choose to use the measurement records for other purposes, perhaps even selling biometric information to other companies. Consider the scenario where Joe, who enjoys leaving his wallet locked safely in his car, has registered his fingerprint to enter and pay for drinks at his favorite dance club. As the club expands to multiple locations, including into other cities, it starts to look for other opportunities for profit. One day while renting a car in another city, Joe is surprised to find that his fingerprint can be used to rent the car. More intriguingly, once set up, the car doors unlock and the car starts only when his fingerprint is used on door and ignition sensors. Suppose Joe next finds that his fingerprint is usable at his favorite restaurant, the athletic club, or for the time clock at his construction job. Some Joes will enjoy the convenience of using their fingerprint at multiple locations and at a variety of stores. Some Janes will be very upset that their fingerprints are being widely shared outside the intended purpose of paying for drinks at the club. How will Joe feel when he finds out that his employer is allowing Web-based sales sites to collect payments for goods and services based on the fingerprint that Joe provided solely to punch the time clock?

Thus far we have only discussed biometric readings that a person would choose to provide to a company, but many biometrics can be measured and captured at a distance, without the subject's awareness. Our opening bar scenario hypothesized a hidden facial recognition system. One's voiceprint might be captured during a telephone service call or help desk call. The bar could probably glean fingerprints from the bar glasses. Someone's gait could be used to identify her as she walks through a mall or a subway station. Her keystroke patterns could be captured while she is using her computer. All of this can be accomplished without a customer's knowledge or consent.

This leads to the first of several recommendations offered throughout this chapter.

> ### Recommendation One
> Companies should carefully explain how biometrics will be collected and used. They should ensure that customers understand and accept the company's plans to use not only biometrics, but the biometric reading in combination with other customer data as well.

Will legitimate, reputable companies collect biometric information and use it without customer knowledge or consent? They already do. Casinos have long employed people, and recently begun implementing electronic systems, to recognize card counters and other patrons whom they would prefer not to serve. These customers are identified mostly through long-range facial recognition.

This presents one of the most interesting dichotomies in biometrics. As personal as biometrics are, the collection of biometric information can be either very personal (like retinal scans and fingerprints) or very remote (like the facial recognition scanning of the crowd proffering tickets at the entrance gates to the Tampa Super Bowl in 2001). Consider that virtually anything about a person can be measured. A friend of one of the authors once started to join the Navy, aiming for the back seat of an F-14. He found that the Navy was concerned with the length of his upper and lower arms, the distance from hip to knee, and his total height and weight. Surprisingly, the Navy had specific ranges for every measurement—all created to allow the ejection seat to work properly. Systems can measure the components of body fluids, the way a person swings her arms when she walks, the distance between her eyes and the size of her ears. Other chapters in this book cover the variety of biometric systems for both voluntary and involuntary collection.

III. Why Is Biometric Information Collected?

Businesses may find a variety of reasons to collect biometric information, including reasons unique to their industry or circumstance. For example, few businesses are without a collection of signatures from customers' checks and correspondence. While we might not think of a signature on a check as a biometric measurement, the bank's implementation of systems pursuant to the Check 21 law[4] means that virtually every check is now imaged, digitized, and available in databases, allowing at least visual comparison of signatures. We should not forget that organizations may choose to collect biometric information from employees as well. Consider, for example, a company that prohibits employees from smoking on company time or property, as well as on their own time.[5] The series of checks made on employees to ensure that they are not smoking is a form of biometric measurement. As the technology grows in power and reduces in costs, businesses may find many problems solvable through biometrics.

However, whether standard identity check or industry-specific analysis, the primary named reasons that businesses and governments collect and use biometric information are the identification of people and the authentication of their identities.

Often, identifying a customer and verifying her identity (also known as authentication) are a prelude to creating a payment for goods or services, the basic building block of business transactions. When we think of people using biometric measures and data, we normally imagine a business trying to confirm its worker's identity for access functions, or its customer's identity for payment functions. We may think about law enforcement using remote biometric identification to find and track criminals. The most basic and best-known uses of the technology fall into these categories.

However, there are many other business reasons, as well as governmental reasons, to collect biometric information and match the information to consumers, citizens, and customers. By its nature, this collected information then becomes a business asset, and subject to uses not initially intended upon collection. Besides identification and verification of users, consider some of the other reasons behind companies' current and future uses of biometrics.

A. Security Influence

Companies may use biometrics as a way to influence their customers. Because biometrics appear to be so scientific and so accurate, people assume many things about the security of systems and buildings protected by biometric identity checks. A simple visit to the bank can provide an example of this security influence of biometrics. The teller is likely to have a simple biometric system used for adding fingerprint identification to checks being cashed—the check casher's thumbprint, captured with ink on the back of the check. Depending on the customer, this simple biometric capture could provide a positive or negative security influence. For example, some people intent on committing check fraud may be discouraged from doing so or may decide to visit less careful banks instead. Meanwhile, law-abiding customers may feel more secure about their accounts because of the visible use of fingerprints to check the identities of customers and to reduce or prevent fraud.

Paper and ink fingerprints, even on the back of a check that eventually is photographed, captured, and rendered digitally, are not the high-tech scanning electronic biometric techniques seen in so many movies or television shows, from the James Bond series to *Charlie's Angels*. Yet paper and ink still perform a service of influencing customer behavior. Consider other ways in which the biometric systems we think about may be used by businesses.

1. Positive Influence

In the days after September 11, 2001, we watched the United States government station National Guard troops at the airports, subject people to more intrusive searches, and generally increase the visible security at airports. Many of these changes remain in place today, especially the most visible shift away from private security firms in airports to U.S. Transportation Security Administration security personnel. In contrast, after the Madrid bombings of March 11, 2004, no similar visible security changes were made at Amtrak stations, and today few or no metal detectors for travelers or x-ray machines for their baggage protect our train passengers.

Visibility is the key to understanding airport security—people can see the workings of the security infrastructure, and they feel more secure as a result. Airport security is directly and indirectly visible to almost the entire population of the United States. Meanwhile, John Kerry created an issue during the 2004 presidential campaign by contending that many less visible airport security measures were being ignored. Discussions during

March 2006 about selling port operations to a Dubai company focused on this dichotomy between the visible security at airports and the invisible (to most Americans) security at ports.

Like our government, companies prefer to create visible security measures. Biometrics, in the form of fingerprints, retinal scans, or other measures, are a visible form of security. Bank vaults have always provided a visible security measure to impress customers. Several banks, as well as other organizations, have adopted biometrics, including picture identification credit cards, as a means of demonstrating a visible commitment to security. This positive perception of security is one reason businesses have adopted, and will continue to adopt, biometric systems. People feel more secure when they perceive a higher level of security.

2. Negative Influence

Biometric systems can also be used to intimidate potential criminals. Whether it is the profile mug shot and ink fingerprints at the police station or the iris-scanning device at employee entrances to many airports, biometric systems can be used to make potential criminals afraid to commit a crime. Criminals tend to gravitate toward places where the profits of their crimes are likely to be high, and where the ease of escaping punishment is also likely to be high. Hair salons, gas stations, and certain convenience stores offer a low profit, but a relatively easy grab and get-away opportunity. Conversely, banks and jewelry stores fit the other end of the spectrum, offering the lure of a big heist, but the deterrence of a vault, cameras, and alarm buttons.

Similarly, armed robbery is risky business (prison, death, or injury are very likely outcomes), whereas check fraud or electronic hacking theft takes considerable physical risk out of the equation. So, why doesn't every criminal move away from physical robbery and toward remote fraud? One reason may be because fraud requires talent, skill, and patience. As a result, many criminals are not convinced that fraud is low-risk or offers a low barrier to entry.

This is where biometric readings enter the picture. Given the choice between a faceless crime and a crime where the criminal must leave behind his fingerprint or other biometric reading, the criminal will generally see the latter type of crime as offering too high a burden and too high a risk. He is likely to choose a less imposing target, one that will not cause him to identify himself. Whether or not a biometric security system improves the chances of catching thieves, it should decrease the chances of thieves attempting to fool the system.

3. Example: Positive or Negative Influence?

Recent visits to theme parks offering roller-coasters and other diversions revealed that these companies are collecting and associating biometrics with tickets. Currently at DisneyWorld, an anonymous magnetic stripe card is associated at first use with a three-dimensional biometric measurement of the ticket holder's first two fingers. Nowhere on the ticket or around the entry area is there any information explaining how these measurements are used, stored, protected, and destroyed. A number of competing theme parks, including Universal Orlando, SeaWorld Adventure Parks, and Paramount Theme Parks, have considered or are considering the implementation of a similar system.[6] One can assume many things about the mechanism, such as the fact that it prevents or reduces sharing of cards by multiple users or that after a ticket has been completely used, the data is destroyed. What other assumptions might consumers make? Is this a positive influencer or a negative influencer?

B. Customer Service

Security is not the only reason for companies to use biometric measurements. Biometrics may also improve customer service. Biometrics may be used to better understand customers, as an additional way to track their Internet surfing and buying habits, or to provide a more personalized customer experience with a business's service.

A business can use biometric measures to uncover several types of valuable information about its customers:

1. *How good a customer can this be?*

Companies may use biometrics as a way to better track the customer and record her visits. Not only can the company tie the user back to payment methods and purchases, but it could also trace customers' movements throughout the store by using gait biometrics (the way someone walks) or face recognition via cameras placed throughout the shop.

2. *How receptive is the customer to my advertising?*

Consider the rewards cards that many companies have introduced. Companies use these cards to tie e-mail, newspaper, and other advertising back to the customer's real spending habits. Unfortunately, companies miss many customers' purchases because customers forget their cards, lose their cards, or otherwise prevent the companies from tying an item purchased to a specific customer. Biometrics capture of identity, particularly for physical store locations, provides an easy way to let customers get their rewards, while allowing companies to gain better data.

3. *What products/services is she most likely to need/buy?*

Companies might use biometrics to suggest products to their customers. People with glasses might be offered contact lenses based on face recognition scanning software. As customer research continues to become more sophisticated, retailers may discover some aspect of face shape, vocal patterns, or stride that indicates that a customer is more likely to buy a certain product, or that she is ready to buy right now. It appears clear that certain behaviors—trust, receptivity to influences, impulse purchasing— may be linked to our genes and DNA. Biometric measures at the DNA level might be used by companies to identify people more susceptible to certain sales techniques and offers.

Clearly this type of profiling slogs into muddy legal and ethical territory. While retailers cannot legally exclude people from their public stores based on a biometric indicator like race or ethnicity, they are free to market African-American hair care products, shark fin, and chicken feet to certain ethnic groups. So where is the line? United States law ties the notion of prohibited exclusion to protected classes of people. Many of these classes are clearly based on biometric characteristics such as race, gender, age, and disability. However, what if a business selects its most promising customers by height or height/weight ratio, favoring certain categories over others? And what if those disfavored categories include a disproportionate number of people in legally protected categories? Is this profiling fair or legal?

What if hormone levels were measured from a woman's breath without her knowledge to identify pregnant customers? What if it turns out that older workers tend to be more impulsive and therefore less well suited for certain jobs; should they be excluded based on their biological propensities? And finally, the entire mess becomes intolerably deep when a business or government uses DNA data to choose the people to target for better prices, harder sales pitches, tax audits, or police searches. The better we become at tying biometric measures to likely behavior patterns, the more intense we make the privacy and fairness problems, and the constitutional issues that follow them.

4. *What are his delivery preferences?*

In 2005, two companies (BioPay and Pay By Touch, which have subsequently merged) began fingerprint-based payment schemes. While promising customers the security and ease-of-use that fingerprint scanners evoke, these companies offer businesses two valuable services. First, the transaction speed is increased. There is no waiting for the customer to pull out a wallet and find the money or credit card, no card-processing time, and no counting change time. With a fingerprint and a bit of processing, the transaction is completed. Second, of course, all the transactions are electronic, meaning fewer hours spent counting (and potentially transporting) physical money. Aside from transaction speed, service delivery is important for other reasons. There are many situations where a wallet or pocketbook is simply inconvenient—from the community swimming pool to the marathon or triathlon registration desk, to the tanning salon. Biometrics for identification, verification, customer service, and service delivery can make sense in many places where traditional forms of identification and payment are inconvenient.

C. Government

Now consider the government's use of biometrics. The most prevalent form of government biometric collection is the huge fingerprint database maintained by the FBI. In addition to the widespread use of fingerprints for law enforcement, new government uses of biometrics are beginning to emerge. At the Tampa Super Bowl and in Virginia Beach, law enforcement officials have tried to use facial recognition systems to identify criminals in crowds. Since September 11, the government has created new requirements (which are still being finalized as of the date of this writing) for biometrics in passports, a set of requirements that affects not only U.S. citizens, but those of other countries seeking to visit the United States.

IV. Current Trends That Will Predict the Future

Laws and regulations tend to lag behind societal and technological changes, rather than to lead them. Therefore, specific laws have not yet been enacted to address the unique challenges of biometric technology and databases or their effects on society. However, information technologies claiming to capture identity have existed for as long as we have been signing our names to business documents. The recent digitization of the names, numbers, signatures, and locations serving to identify us has led to a heightened level of concern over the ability to gather, aggregate, and steal this information from any connected terminal in the world. The laws and regulations, both passed and proposed, addressing this recent concern serve as a reasonable road map for projecting into the uncharted territory before us.

A. Privacy Protection under the Current Law

Since the rise of the Internet as a significant business and governmental tool and the resulting interconnectivity of companies and consumers, the government has tried to address the privacy of significant personal information. The European privacy regime is currently the most strict, as European governments treat privacy of certain facts and information as a natural right held by individuals and enforced by the state.[7] The belief that individuals have enforceable rights to keep certain information private changes a society's entire approach to electronic information management. If individuals are the unequivocal owners of their information, then those businesses or governments gaining access to that information are essentially holding it in trust for its owners, and must receive specific permission from the owners to use the information. Thus, generally speaking, European businesses and government must request permission from consumers to use the consumer's information for any purpose other than its original intended use,[8] and otherwise are responsible for protecting and limiting the use of such information.[9]

While it is currently unsettled how laws in the European Union would treat biometric samples taken for the purpose of identification or authentication, it is clear that such samples will belong in essence to the person who offered the underlying biometric sample, not the business who took the sample. That business will be restricted from selling or publicizing that biometric indicator to identify an individual without the individual's express permission. The burden will be on the business collecting the biometric data to refrain from using the data for any purpose outside of the identification function originally intended.

B. American Privacy Model

Conversely, under the American privacy protection model, businesses and governments are assumed to own the information they collect from and about individuals, but certain information that can used to identify individuals is protected from further dissemination without the express consent of that individual.[10] Personal privacy, if a right at all, is not a very well-defined right under American law.[11] A right to privacy is not mentioned in the United States Constitution.[12] Instead, as described below, the United States has recently erected a series of piecemeal restrictions on use that impose obligations on business and government to treat the personal information of others more carefully.

Legislatures have constructed special legal protections around data relating to children, to a person's finances, and to a person's health or health care.[13] In cases where a business accepts personal information under a publicized privacy regime, that business must comply with its own stated rules.[14] However, where personal data falls outside these categories, and a collecting company has not committed to keep such data private, the company can gather the data and sell it to the highest bidder, use it for the most nefarious of purposes, or use it to enhance the customer's next experience with the company, with or without the customer's permission. Businesses collect information relating to their customers' purchases, preferences, and every contact between that customer and the business. While this information tends to be protected dearly as a trade secret of the business that collected it, much of the information is not protected from release or dissemination under American law. Furthermore, when a company declares bankruptcy, the customer data collected over years of transactions may be the only valuable asset the company owns—the only asset a bankruptcy trustee may sell to satisfy creditors.

As of this writing, no United States jurisdiction classifies biometric information as a protected class, except to the extent that such information may reflect on the health or the health care of the person who gave the information. If a fingerprint or retinal scan or gait analysis is used as an identifier, no specialized restriction prevents the holder from giving or selling the biometric data to others.[15] There may theoretically be restrictions under HIPAA on the harvesting of DNA from a person, and using that DNA to draw health-related conclusions about the donor; however, as of this writing, the statute has not been interpreted in such a fashion. In short, under the laws of the United States and from a legal business perspective, biometric samples used for identification or authentication are not currently considered to be private data belonging to the donor.

If, or more likely when, the nearly instantaneous DNA identification test in the movie *Gattaca* were possible, American-based companies collecting such DNA would effectively own and could sell the DNA data. As we learn more about the relationship between DNA and disease or behavior, the value of this information will increase significantly.

C. Privacy and Security Practices

Privacy laws address a business's or government's ability to use personal information in a manner beyond its original intended use. However, privacy laws cannot exist without the security promise implied in any such use. In other words, a business may violate its responsibility under privacy laws by selling the private information to a third party or by using the information in some unauthorized fashion, but it may also violate its responsibility by failing to adequately protect the information and allowing another person to steal it. A dry cleaner is not allowed to sell his customer's clothes out the back door of the shop or to wear his customer's clothes out to a party; similarly, he will be held liable for leaving the back door open and the shop unattended so that thieves may take the customer's property.[16] The flip side of privacy protections is information security requirements effecting the protections.

Information security responsibilities have recently solidified into a robust series of standards driven by statute, by regulation, and by case law. The financial service industry has a long history with security requirements, which have grown with various business technologies over time.[17] At one point, a bank's security was mostly walls, bars, and vaults. Now, Internet firewalls, encryption, and secure access points with password protection rule the industry, and the government's expectations regarding protection of the digitized treasure within the bank computers demonstrate an acknowledgment of this technological shift. Current financial regulations were formed following observation of the evolving networked, digital world.[18] This is a world that continues to evolve faster than law and regulation can truly keep pace.

On the other hand, protection of health-care information arrived in one fell swoop in the form of the HIPAA law and regulations, which specifically address information security as part of an overall health data protection scheme.[19] These regulations require that companies collecting and keeping personal health-related data must install adequate policies, tools, and procedures to ensure that such data is not inadvertently exposed.

Child protection statutes have not been written with information security standards,[20] but law enforcement officials at different levels have been willing in the past five years to sue to hold businesses responsible for maintaining the safety of protected information.[21] For an organization with information on children, information security standards have arisen from cases in which a company's security policies were found to be poorly constructed. Therefore, companies are not instructed on the minimum acceptable security practices, but must guess the security standards by learning which specific practices were considered by the courts to be insufficient. This results in a risk that companies must manage. Furthermore, operating blind in this fashion is the norm for most information security practices in most industries.

Finally, a new form of information security law (S.B. 1386) has been implemented in California, requiring businesses to inform their customers if that customer's information has been compromised by potential thieves and/or computer hackers.[22] This was the first statute in the United States that addressed the problems of data privacy and security by requiring customer notification of potential security breaches. As news of cascading data compromises and information security violations crescendos in the American consciousness, many other American states are now considering notification laws of their own.[23] The Office of the Comptroller of the Currency (OCC) recently began requiring all financial institutions to follow the California notification law as well.[24] These laws mark a real change for the information security community, which has traditionally operated in darkness. Security breaches have generally been internal matters that the institution addresses without bothering to tell its customers. On some occasions those companies involved have shared information about breaches only haphazardly among tiny circles of information-security professionals. In the new era of customer notification, the rules have changed, and openness is replacing paternalistic silence.

There are many possible results from this new openness. It is possible that one effect of this new open policy will be a reduction of the public's trust and confidence in the financial services sector stemming from the public's realization that the sector is not a monolithic secure fortress on the hill. However, there has been little evidence to date of bank customers withdrawing their money to hide under a mattress or to bury in a mayonnaise jar in the back yard. Another possible development is comparison shopping for the "most secure financial institution." The trend toward laws requiring more openness will certainly affect the treatment of biometric identifiers and the security of the corporate and governmental databases that hold them.

D. Law Enforcement and the Constitution

Under the law of the United States, the government is subject to additional responsibilities and restrictions concerning the manner in which it collects, holds, and uses information on its citizens. Many of those restrictions are written into the Constitution,[25] and many others are statutory.[26] Since the passage of the USA-PATRIOT Act, some of those restrictions have been loosened in the fight against terrorism, but the government, and especially law enforcement, will always have more and different restrictions on its information-gathering than businesses or other individuals might have. Law enforcement agencies needing to prevent terrorists and other suspects from learning of their suspicions are generally well prepared to protect private information. These are the tradeoffs our free society has made around protecting personal information. However, they do not preclude the use of any type of biometric information for identification or as evidence in criminal cases. The constitutional restrictions instead may affect the permissions needed to collect and use certain biometric information.

Earlier in this chapter we made a distinction between private biometric information, which is invasive or intrusive to the individual, and which an individual may choose to give or to withhold from the collecting business or agency, and public biometric information. Examples of private biometric information would include a retinal scan, fingerprints, DNA, blood type, and even a written signature, all of which generally must be given with the consent or assistance of the individual subject. Contrast this private biometric information with public biometric indicators like face geometry, voiceprints, and gait pattern that can be collected from an individual without her consent or even her knowledge.

The government needs a warrant to collect private biometric information. Current restrictions on government action in investigating its citizens led to the need for government to follow certain rules—for example, make a proper arrest, obtain and serve a proper warrant, or obtain a specialized court order—in order to be able to take the private biometric information from a citizen. However, the government, like businesses, currently is not subject to any restrictions on the taking of public biometric information. Current U.S. laws assume that a person introduces his voice, his face, and his gait to the world at large when he chooses to use them in an open forum; therefore, the government can take impressions of these biometric indicators without infringing upon a citizen's privacy. Public biometric information can be collected in any public space by anyone.

E. Proposals Affecting Biometrics

While no law specifically regulating biometric identifiers has been signed, nothing has prevented legislators from trying to address the issue.[27] Privacy advocates believe that the maintenance of biometric databases would be dangerous for consumers and could harm individual rights. They see a slippery slope descending from the first permissible biometric use for business to a national biometric identity system that would strip away our individual privacy, as predicted in Orwell's *1984* and Spielberg's *Minority Report*.

If enacted, legislative proposals restricting biometric data storage would throw the entire industry into chaos. Businesses spend significant money building identification systems. Identity management has become a significant subcategory within the larger information security space. Businesses are now beginning to invest in biometric-based identity management systems, and limiting their use in some jurisdictions would create huge investment requirements with little reward for businesses. Restricting the storage and linkage of some data, but not other data, would create a multiclass system, with customers receiving different service based on where they lived (or where the system thinks they live) rather than on their customer status or, more important, business requirements. Despite the chaos of multiple jurisdictional rules, privacy advocates and others are likely to win some legislative battles, especially if the biometrics industry is unable to better contain the leaks of personal information and debit and credit card data revealed during the last two years.

> ### Recommendation Two
> Because the future is uncertain, our second major recommendation is that companies plan to carefully manage any private biometric data they collect from employees or customers. A well-designed system, supporting a variety of rules and options will be costlier up front, but will provide companies with the best protection against uncertain biometric rules.

F. Business and Biometrics

We have identified numerous overt applications of biometrics in everyday life:

- Payment at supermarkets, sandwich shops, convenience stores, and coffee shops
- Access to bank safety deposit boxes
- Airline/TSA retinal scans for airport access
- Time clocks where sign-in and sign-out is by fingerprint instead of timecard
- Virginia Beach and Tampa Bay face-recognition projects

Consider the impact as these biometric applications result in the collection of more and more information by more and more businesses. For example, is the legal system prepared to address such a change?

G. Is the Legal System Ready?

As noted, the law lags behind technology, so the legal system may not be ready for widespread biometric systems. Laws and regulations in the present system are beginning to anticipate the changes in society, technology, and business that may be wrought by widespread use of biometrics. We are just testing the limits of intrusions into privacy and the allocation of responsibility for breaches of security. As many of the current laws and regulations in this area have only been in force during the past decade or less, American society has not taken time to reflect on whether these approaches will truly protect society's interests. In addition, some of the key issues sure to spark the debate about widespread use of biometric identifiers have not been addressed at all by current laws.

For example, the election of 2004 demonstrated the power and hinted at the danger of mega-databases. Companies like ChoicePoint, Axciom, and others have vast databases gathered and maintained from public and private records. Like companies seeking consumers, both Democrats and Republicans used these databases of aggregated personal information to target specific groups of likely voters. The political parties sliced and diced the data to find new, relevant interest groups that could swing a close election. A fight over "Soccer Moms" and "Nascar Dads" in previous elections had been honed in a sophisticated fashion to instead target "churchgoing men under 50 with families and SUVs who subscribe to sports magazines and browse political Web sites living in the zip codes outside of the first suburban ring of Kansas City and Saint Louis." These collection and analysis firms aggregate thousands of public and private facts about each American consumer/customer/voter. They can break down each person's life into purchasing preferences and geographical/anthropological assumptions and then sell the information to the highest bidder, along with the home addresses, telephone numbers, and e-mail addresses of each target. This allows a highly tailored message to reach its intended, receptive target.

This kind of narrow and well-defined data mining is perhaps exemplified by the *Chicago Tribune* investigation revealing the identities of CIA agents.[28] Imagine the additional opportunities based on finger sizes, hair color, and other biometric data that could be mined.

Through data aggregators, on-line and off-line data can be merged into one comprehensive portrait of a target to be approached individually or carved into a specific lifestyle-based interest group. In 2005, ChoicePoint announced that questionable businesses had fraudulently gained access to its database, perusing the database at will.[29] This access allowed the attackers to steal valuable information on thousands, maybe millions, of Americans. Yet, no public outcry produced any law limiting the aggregation of this personal information. Some of the information in these databases may have been illegally obtained, illegally disseminated, or used for improper purposes. But the actual act of creating this kind of aggregated database is permitted under law, and there appears to be no likelihood that any legislative or regulatory body will address the building and maintenance of such information collections, searchable by characteristic or by the individual person.

Moreover, current regulations do not address standardization in most technologies. As will be discussed in the next section, the U.S. government has been, and will likely continue to be, disinclined to directly impose standards to govern biometric technology. However, it is possible that a government-driven biometric program, like a fingerprint identification system tied to passport control, could arise as a de facto standard for industries wanting to tie into the government's investment in the technology and the likely availability of cost-effective biometric readers that would arise from such a commitment. This is the influence of government as a large consumer of the product, and not the dictation of government as the highest regulating body. The latter is simply out of character for the current role of regulators and legislators in the technology marketplace, and that is unlikely to change in the near future.

Finally, very few types of information have been guarded by government action. For example, the state is generally quick to express its interest in protecting children. Similarly, financial information has long been the subject of protective regulation. Video rental data is covered under federal protections.[30] Only recently has data related to personal health care been sheltered by government action. However, many significant aspects of our lives generate information. Our housing and automobile purchases are a matter of public record. The purchase of food, sundries, over-the-counter medications, books, magazines, furniture, electronics, business items, and nearly all services may be recorded and the transactional information sold to anyone. The fact that a person has ordered and watched certain shows on his cable and satellite services can be recorded and passed to anyone. Visits to Internet sites are generally monitored and such activity can be collected in a file attributable to the computer user as Web site owners share information and mine aggregators to combine little bits of information into the mosaic of the user.

Will the United States government choose to protect biometric identifiers in the same way that it protects medical information? In some ways, the two types of data are similar, being based on information about an individual's physical being. Or will the government treat biometric data in the same way it treats most information in an American's life, that is, with benign neglect, and offer it only rare protection? No current biometric data is sheltered by law from disclosure, but there may be reason to believe that the recent movement in California to ban collection of biometric data may indicate the legislature's aims to treat biometric data as an additional class of information entitled to special protection.

V. Living on the Fault Line—Biometric Issues That May Arise

Given our review of biometric technology, its uses in business and government, and the current legal regime, we can begin to examine the likely effects of widespread utilization of this technology. Assuming widespread use of biometric identification and authentication technologies, we can see many controversies. There are controversies brewing over standardization of the many forms of collecting biometric information. There are conflicts between technologies for capturing and storing biometric information. As more ways of measuring people are invoked, particularly without explicit permission, new concerns about surreptitious collection will arise. Further, privacy, already a concern to many, will become an even more significant public issue as people worry about surveillance. We can see problems arising with the ever-expanding universe of super-aggregated databases combining biometric information about a person, including the person's spending, worshipping, and voting habits. We perceive issues with registration, government-issued biometric identity cards, and interpretation of biometric data. The technology measuring our physical beings will create legal and societal firestorms pitting a business against its own customers and the government against its citizens.

A. Acceptance of Biometric Standards

One issue that may help or hinder the growth of biometrics in the work-place, and will most certainly cause controversy in the near future, is the prospect of choosing technology and policy standards to implement a bio-metric identification or authentication regime. As biometric identifiers be-come more widely used by business and government, each organization will have to choose which biometric indicator to measure, how to gather the information, how to measure the information, how to distinguish one set of information from the next, how to store the information, and how to use the information once a database is built. For example, if a bank chooses to use iris scanning for access to safety deposit boxes, bank management must decide when such scans will be required and what type and brand of scanner will be used to take the biometric information from customers or employees. The bank will need to decide the level of detail measured on each iris—will the bank accept and store a full picture of the iris or simply measure against certain parts of the eye pattern? How many points of the pattern will be measured, and how many points of comparison in the iris will be accepted as being sufficient for a positive match? Does the bank store the data in some standardized format that can be passed to other affiliated companies or used by the government? Does the bank choose a familiar off-the-shelf platform from some "Microsoft of biometrics," or does it instead use a proprietary system that will only work for this one function? All of these questions will apply to the implementation of any biometric business or government system, and they all involve choosing a system of standards.

The most organized—and therefore the least likely—method of choos-ing a technological standard is a group of interested industrial and govern-mental participants teaming together to pick a standard that all of them will use. Technology standardization of this type occasionally occurs. For ex-amples, see the wireless security standards or the public key encryption system. But frequently, the competing interests of industry and govern-ment participants conflict too greatly to make agreement possible. Given the complexity of biometric security, it is possible that certain aspects of the practical technology, like data storage or fingerprint mapping, could develop as an agreed standard, but it is highly unlikely that an entire set of technologies for capturing, holding, measuring, comparing, and disposing of such data would be decreed by a general agreement of all involved parties acting on behalf of the public interest.

It is also unlikely that a technological standard for the creation of biometric data will be consciously chosen by the United States federal government. Recent administrations have avoided choosing winners and losers in technology formats, from electronic encryption to fuel-efficient engines. Governmental bodies have shown a preference for setting goals and performance requirements and allowing the market to decide which technology best meets its own requirements. For example, the government did not step in and declare that the Sony Betamax videotape recording format was superior and must be used for all videocassette recorders. Instead, the market eventually decided to prefer the VHS format, which became a de facto standard. Except for military equipment, public transportation issues, and highly regulated industries like radio (where the government allocates the spectrum) and telephones (where a government-enforced monopoly was granted to AT&T and later taken away), the U.S. government does not usually make choices on technology standards for its citizens. This allows market-grown monopolies, like the Microsoft personal computer operating system monopoly, to arise, or industries like wireless Internet access to flounder as no standard emerges. We can assume the same will likely be true in the industries that arise around biometric identification, authentication, storage, and application. The U.S. government is unlikely to impose a standard or group of standards that would drive industry choices.

Instead, as biometric systems expand in our society, it is likely that some standards will not be chosen by all participants or imposed by the government but will emerge from common usage. If, for example, the Homeland Security Department decides that full-fingerprint pictures are required for any person entering or leaving the country, then an enormous and influential database will be created. First, law enforcement and other departments of the federal and state governments will begin using the same database standards for their own purposes, similar to the creeping governmental use of an individual's Social Security number for taxes, benefit distributions, and other methods of identification. Next, businesses will begin to build their own systems around the already-existing governmental data. In addition, whatever company wins the contract to build such a huge and important system will likely be able to parlay that experience into system design for other clients. Hence, a de facto standard is born. One large and powerful entity chooses it and pays for it. The rest hop on board and ride the same standards.

The legislative requirement for biometric passports domestically and internationally will clearly drive a small set of standards, or a single standard, for passports established by the international passport community. This is likely to become a de facto standard for at least the form of biometric (likely fingerprints) chosen by the community. This choice will be heavily influenced by the U.S. federal government's technical standards bodies, the Department of Homeland Security (DHS), and the National Institute of Standards in Technology (NIST). Knowing that the Social Security number was adopted as a de facto national identity number, one can predict that the passport standard will become a de facto international biometric identity. The International Civil Aviation Organization (ICAO, a United Nations body) has set the standards for passports to include biometrics. The ICAO requires an electronic, stored copy of the passport holder's picture, and allows two additional biometrics—fingerprints and iris scans.[31]

1. *De Facto Standards versus Chosen Standards*

As noted, de facto and chosen standards around biometrics are likely to arise. Mostly, real standards will be combinations of de facto standards, with banks and others adopting the standards chosen by the government.

De facto standards may arise by other methods. If a large technology provider like Microsoft or IBM decides to move seriously into biometric identification or security, then the market may accept the standards that the vendor has chosen. IBM is currently offering fingerprint-scanning hardware on certain high-end laptops, and, if the trend spreads to other company products, a de facto standard for fingerprint identification may be born. De facto standards can also arise around an application or legal requirement. Where a consensus forms that voiceprint scanning is required for identification of customers seeking telephone support, then it is likely that a few, and maybe one, set of technologies and measurements would be applied to that problem by all companies that feel compelled to use it. Similarly, where the law imposed the obligation for significantly better security than currently exists in certain business situations, then industries may choose to rally around a biometric standard to meet this obligation. Currently, as noted above, information security legal requirements are growing. Businesses are affected by laws and regulations like HIPAA, and by standards arising from case law when a certain security practice is found to be inadequate. If cases or regulations in one industry—banking, health care or child care being most likely—require a substantial improvement in basic security, and biometric measuring technologies both fit the problem and could be demonstrated as cost-effective, then the industry would likely need to move quickly to implement the biometric technology. The safest, fastest, and least expensive way to do so would be to agree on technical standards that the entire industry could use and that the government could verify. We would see de facto standards by legal fiat.

B. Evidentiary and Procedural Issues

Courts have used fingerprints to prove identity in courts for the last century. Too often, we find ourselves watching courtroom dramas where the case is settled because of the certainty of fingerprints. Interestingly, fingerprints were in the news in 2005, associated with the FBI targeting an Oregon lawyer for a role in the Madrid train bombings.[32] The FBI was so convinced that it had identified and captured the culprit after their computerized fingerprint system identified him, and three experts reviewed the match, that, rather than believe Spanish authorities, the FBI sent their own team to Spain before the lawyer was cleared.

So if full fingerprints can fail to accurately identify someone, how well can biometric systems provide evidence beyond a reasonable doubt? The answer depends on the system and specific methods for matching biometrics. However, in most cases, systems will have false positives or false negatives on the order of one in 1,000 to one in 100,000. Few systems possess one in 1 million quality. This is why the cost of implementing fingerprint systems for passports seems so high.

Facial-recognition systems work only if the photograph is taken with proper lighting and an especially bland expression on the face. Even then, the error rate for facial-recognition software has proved to be as high as 10% in tests. If that were translated into reality, one person in 10 would need to be pulled aside for extra screening. Fingerprint and iris-recognition technology have significant error rates, too. Therefore, despite the belief that biometrics will make crossing a border more efficient and secure, it could well have the opposite effect, as false alarms become the norm.

Assuming a modest 10 million visitors per year, a system with a 1 in 10,000 error rate would still create 1,000 errors a year. The United States receives hundreds of millions of tourists each year; this would result in tens to hundreds of thousands of errors that must be dealt with.

As with any scientific process, biometric readings provide an illusion of infallibility that may be difficult to overcome in court. If a jury sees that a fingerprint was analyzed by the biometric reader, jurors will likely assume it was read correctly. In court, what constitutes a reasonable doubt for the reading of a fingerprint? One in a thousand? One in a million? Juries will have to continue adjusting expectations of proof and doubt as biometric measurements change over time.

C. Database Issues

In most cases, individuals concerned with privacy express their dismay over the vast stores of information that companies now maintain about each person. Numerous articles describe the early fears concerning Social Security numbers. Politicians promised that Social Security would not become a national identity system. For the government, that is often true. However, even the government uses Social Security numbers widely, especially for taxable and taxpayer-sponsored benefits. The most obvious of these is the Internal Revenue Service, which requires infants to have a Social Security number before they can be counted for tax deductions. Of course, the IRS will use that Social Security number to track dispersal of a person's estate to his survivors, creating a straightforward National Identity system from birth through death.

The private sector is far different. Data aggregation companies, the credit bureaus, and others work to understand their customers, and the best way to sort customers is often by Social Security number. When combined with name and address, the Social Security number is nearly perfect at identifying an individual in the databases. Using the Social Security number as a reference, companies can combine bits from various sources into a more complete picture of the individual.

Most likely, companies will use a biometric identification to tie back to a person's name, address, and that all-important Social Security number that is already widely used to index and link data about a person in one database with data in another. We describe above how meta-databases currently operate and how their use may be projected into biometric applications. However, the data collection can become more complicated as government intervenes

1. Metabases—Government Participation

Let us assume that Fred's Bank collects financial data on its customers, and Larry's Widgets collects biometric and transactional information from its customers, while MetaID Base is a company that aggregates many kinds of information about people's lives and identities. Suppose that the government has become interested in people who have bank accounts and use Larry's Widgets' widgets on a regular basis. (Perhaps such people are likely to be or become terrorists based on recent terrorist activities.) Depending on the specific agreements at our hypothetical companies, the government might have to approach all three in order to find these potential terrorists. However, it is likely that the government can simply mine the data at MetaIDBase. Using these meta-databases, law enforcement can gain indirect access to information about people without approaching either the people or the companies they do business with. After all, Fred's Bank customers have no relationship with MetaIDBase at all. They may not even know MetaIDBase exists.

There are many concerns with this system. Two concerns are addressed in the next section in more detail. First, we have discussed the bleeding of information used for one purpose (to access a bank account at Fred's Bank) into another purpose (to ensure the accuracy of Larry's Widgets' database(s)). The next section expands on the bleeding concern. The second issue is that customers have no way to identify and address concerns with incorrect data at MetaIDBase.

2. Bleeding Data—Centralized versus Distributed Database

Companies can always store biometric data in centralized databases, just like they store customer information such as names, addresses, etc. Anticipating privacy and other concerns, the makers of these systems have used other technologies to create systems that store the biometric data in a secure container, like a smart card. In these systems, the biometric data is not centralized but is distributed, perhaps with the only copy stored in a card the bearer keeps and controls. A registration system and a well-designed storage device can allow companies to gain the benefits of biometrics, while allowing customers the sense of privacy derived from not leaving their fingerprints in the company database.

Generally, companies use centralized databases, leading to privacy concerns, because the database is connected to networks like the Internet. A third option is for companies to use a local storage system. For example, access to a safety deposit box in a bank branch would use a local storage system (the only safety deposit boxes a customer can access are in that one branch). It makes little sense to centrally store all of the biometrics associated with the safety deposit boxes of many branches, because whatever is stored in the box is only in that one branch.

Companies need to consider their system and balance the pros, cons, and overall risks associated with centralized, distributed, and local storage. They can implement a system that meets their needs and acceptable risk levels.

3. Non-permissive Bleeding

There are so many databases with information about people, it may be difficult to maintain all of them. People move, die, marry, divorce, and procreate. In the process, databases receive conflicting information. As an example, consider the FBI fingerprint database, which effectively captures the fingerprints of all criminal suspects. In May 2005, *The New York Times* noted that the FBI apologized for failing to identify a murder suspect.[33] The suspect had managed to have his information entered in the database twice, once in his real name, and a second entry under an assumed name. The FBI system was not prepared for this data-entry error, and allowed two identities with the same fingerprints, failing to compare new entries to those already in the database.

As a company builds out a system, it should remember that all the data may not be correct. System components can fail, hardware wears out, and other data entry may be erroneous as well. All systems need a method for consumers, customers, and others to fix erroneous data. Database errors will continue to affect information systems, and these errors will lead to more drastic consequences as people are mistaken for others by computers with biometric information.

D. Growth of Legal Protections

Current United States law is beginning to protect privacy and to recognize information security requirements. Only in the past 10 years has this country formed the start of a mature recognition of information security, authentication, identification, non-repudiation, encryption, and the obligations surrounding them. It is highly likely that regular use and storage of biometric data would further advance the law and regulations around privacy and security.

We see now that some jurisdictions fear any use of biometrics, as attested by the regular introduction of bills in the California legislature to ban any use of biometric data in business and government. As more databases are breached, as more identities are stolen, as the secondary market for personal information grows, so will this fear. One strand of thought and emotion surrounding the privacy of biological data (data that, unlike passwords or assigned numbers, cannot be easily changed) is fear of loss or intentional misuse of the data. There is the fear of truly losing one's identity because biometric data marks a person in a most physical sense. To register a false negative for biometric data would be to deny a person his own being, while to register a false positive would be to award a nearly undeniable false identity to the crook who fooled the system. Such a fear is too primal and too rational to be ignored, and it may expose itself through limits on certain uses of biometric registrations.

While the supreme usefulness of biometric recognition systems makes it unlikely that they will be banned entirely, it makes sense that certain uses of biometric measurements or procedures surrounding their use might be restricted. A probable example would be restrictions on forced production of private biometric information. As we write this article, certain states have limitations on when and where a person can be forced to divulge her Social Security number in a commercial setting. A retail store may wish to take the customer's Social Security number to verify identity when cashing a check, but in most instances, the store cannot require the check writer to produce a Social Security number in order for the check to be accepted as payment.

Similarly, future legislatures, foiled in their attempts to ban biometric information systems outright, may decide to take a piecemeal approach and find a lowest common denominator that will garner enough votes to restrict such systems in a limited fashion, banning stores from requiring biometric signatures from its patrons. These limitations are likely to result in a balancing of biometric interests, whereby employers and law enforcement are able to force people into divulging biometric information, while retailers will have to pay for the privilege with discounts or other "rewards."

This leaves wide open the retail use of "public" biometric information. If the local grocery store is precluded from forcing a person to reveal his fingerprint to confirm his identity, will it be allowed to use a face-geometry scan to serve the same function? Those privacy advocates who fear misuse of all biometrics may not perceive a major difference between the danger of collecting and storing private biometric data and the dangers of collecting and storing public biometric data. However, a scan of a person's face, gait, or vein patterns may prove to be impossible to regulate. So, it is most probable that private biometric data may be protected, while public biometric information resists all attempts at regulation. The argument that the "display of your public face comes with an implied right to use it to identify you" is likely to prevail. This argument is similar to the prevailing legal Internet position that, with few limited exceptions, posting a site on the public Internet includes an implied right to link to the site.

Conversely, we have also seen a trend toward requiring businesses and governments to produce higher levels of security for certain functions, and biometric systems provide one of the best forms of identification and authentication available to meet those security requirements. As the prices for biometric systems come down and the need for security against terrorism, fraud and abuse grows, it is likely that certain biometric authentications could be recommended or required by law. Currently, certain businesses are obligated to use at least "state of the art" security measures to protect valuable information and physical access to secure areas. When biometric systems are seen as the state of the art for identity confirmation, the law will require their use for these same protections. It is unlikely that biometric matching systems will be specifically required by state or federal statute, due to legislators' fears of biometric indicator theft or misuse. The requirement of biometric systems in the United States is more likely to arise from court decisions or from regulations recognizing the state of the art in authentication and requiring that businesses keep up with current technology. In other words, these protections and requirements for using biometric systems are likely to arrive through the back door.

> ### Recommendation Three
> Stay abreast of changes in the law, especially if your company is collecting and using consumer biometric information. Also, consider participation in legislative and regulatory policies as they affect your industry, including voluntary standards.

E. Government ID Cards/Default Cards

The U.S. government's efforts to fight terrorism are already increasing the use of biometrics for identification and authentication. Well before 2001, the Department of Defense (DOD) Common Access Card (CAC) was widely deployed. Although lacking a biometric measure, the card provided a secure operating system and storage of the user's private key, and it provided the basis for a standardized ID card for the military. Following September 11, 2001, the government began a series of activities intended to move the CAC program into the larger government ID space. The President signed Homeland Security Decision Directive 12, which addresses the standardization of government identity cards. The card directive led to additional programs, some involving the collection of biometrics.

F. Collection Issues/Registration

There are a variety of issues facing a company that has made the decision to use biometrics. In most systems, registering users is the critical first step. How does a company ensure that the registrant is the person that she claims to be? If Joe registers as Fred, then the company loses its ability to know what Fred is really doing, whether that applies to Fred's preferences in company products or to Fred's use of company services. Depending on the company and system, it may mean that Joe can impersonate Fred, perhaps moving money, stealing tangible or nontangible assets, or otherwise creating havoc.

Depending on the importance and value of transactions associated with the biometric methods, the company may need to be more or less controlling in the processes associated with registering users and collecting the initial, stored value of a biometric. For a business such as a tanning salon, registering the holder of a credit card may be sufficient identification. After all, stealing access to a tanning bed is neither especially valuable nor something one can get away with forever. In contrast, access to top-secret, secure rooms may require a registration process that prevents a single person from gaining access, forcing collusion between multiple people. Such a registration process might capture all 10 fingerprints for long-term storage and require a birth certificate, a driver's license, and a background check. For access to a corporate banking trading floor, something in between these two extremes might be required.

Because so much data may be collected, including logs about the registration process activities, companies should consider the security of their data, database systems, and data collection methods. The Electronic Privacy Information Center has several interesting resources on biometrics, suggesting six areas that need to be considered carefully to ensure consumer privacy for biometric systems.[34] These areas include:

a) Storage of the data
b) Vulnerability of the data to theft
c) Confidence in the technology to identify users
d) Authenticity of the users and registration process
e) Linking of the biometric data to other databases
f) Ubiquity of the information trail about users

These are all important areas and are worthy of broader consideration, but are outside of the scope of this chapter.

Companies trying to manage biometric collection issues need a legal and ethical starting point. Many organizations, from the Better Business Bureau[35] to the Direct Marketing Association[36] to the Electronic Privacy Information Center,[37] have documented important components of privacy policies and practices. These guidelines often put a high value on ethical company behavior, and many regard privacy of customer data as the highest goal. Customer privacy may limit the company's ability to use the data, so companies may want to choose more company-friendly policies at the risk of offending potential customers. Nonetheless, these customer-first privacy guidelines are a good starting point because they provide the highest degree of confidence to the consumer.

Of course, companies may also wish to take a more company-friendly perspective on biometric data, collecting it by default and forcing the customer to opt-out of the system. Companies should consider their approach carefully and strategically, because changing from a default opt-out to an opt-in system can be costly and time-consuming.

Typically these privacy recommendations take into consideration several aspects of customer data and privacy for that data. While data security is well understood, if not well practiced, privacy and data collection practices in the United States have been biased toward the advantage of companies. While companies may continue existing practices with regard to adding biometrics to their data collections, consumers may be put off by giving up more of their personal information—arguably their most personal information—without adequate safeguards.

Recommendation Four

The following aspects are worthy of consideration by companies in the setup of their biometric policies, practices, and systems:

a) *Notification:* Notification to the customer of the collection of the information, of the type of information collected, and the ways in which the information will be used.

b) *Privacy Policy:* Explanation of how the company will protect the customer's information.

c) *Request for Consent or Opt-In:* Allowing the customer to choose whether to allow the company to collect and use the information.

d) *Reminders:* Periodic reminders to the customer of the policy and the information stored and notification of any changes in policy.

e) *Opportunities:* Opportunities for the customer to review and correct the information being held by the company.

f) *Repeated Consents:* Periodic requests for the customer's permission to maintain the information.

g) *Extension:* Requests to the customer to opt-in to any extended or additional uses of the collected information.

Consider a company whose legal counsel recommended following a customer-first perspective. The company decided to follow an opt-in strategy, providing good customer notification of its planned biometric systems and usage. Ultimately, all seven aspects would be considered as the company sets up its biometric system.

1. Notification

One of the early and important aspects of collecting biometric or other personally identifiable information is notifying people that data is being collected. For most biometric systems, the customer or user is an active participant in providing the data for collection. Usually this occurs during the registration process. However, notifications may not always be made or may be indirect.

As an example, consider a typical company policy governing the use of corporate communications systems. Often a company will say that users consent to monitoring. The types of monitoring, the types of data collected, and the ways in which it will be used may or may not be explained.

Now consider a gait-based biometric system used in a public train station. People visiting the train station are most likely not told that they are being watched, although in more and more cases, organizations are notifying customers that video recording may be used. At the train station, customers may assume that they are being watched. There are security guards and both visible and hidden cameras. However, there are no expectations by such customers that their walking patterns will be observed, catalogued, and collected. Thus, when the train station starts identifying customers by their gait, perhaps to provide better customer service or to find customers who use the train regularly, will the customers have been notified? Will companies collect biometrics without notification? As illustrated earlier by the examples of the casinos, the Tampa Super Bowl, and Virginia Beach, they already do.

Notification should be clear, so that the customer recognizes that he has been notified. Notification should also include several pieces of information: what data is being collected, how it will be used, and how it will be protected (e.g., a privacy policy). Records of notification should be maintained.

2. Privacy Policy

Any organization collecting personally identifiable information (names, Social Security numbers, credit or bank account numbers, addresses, and similar data) should have a privacy policy. Privacy policies also cover what kinds of information the organization collects, how the information is valued and protected by the organization, and what the organization will do with the information. Collection of certain types of information, like personally identifiable health information, requires a privacy policy that is at least as protective as the relevant statutes. However, a company may create serious problems for itself by publicizing a privacy policy that does not accurately state the company's data privacy practices, even where that company had no obligation to make or keep the data private. Where a company creates a privacy policy, the company must follow the rules or be held to task by the government for breaking its promises.

3. Request for Consent or Opt-In

As part of the acknowledgment of the notification, companies should consider requesting the customer's consent. Studies show that simple "click through" acknowledgments are virtually ignored by most people. While such acknowledgments meet the requirements for customers opting into a system or consenting to the capture and use of their information, they may leave customers confused. Two recent computer security issues, adware and spyware,[38] often trick users into accepting them by misleading the user or by providing no opportunity for the user to understand and accept the malicious software.

A related acknowledgment concept is called opt-in. In the United States, where companies regularly maintain data about users, consumers have to work through a process of opting-out of the organization's systems. The National Do Not Call List is probably the most well-known opt-out system, offering users the chance to opt-out of being disturbed during dinner by certain marketing calls. For biometric systems, companies should consider the possibility that customers will want options other than biometrics, and certainly that some will not want to provide biometrics.

Consumers will face a variety of choices from organizations with respect to biometrics. In some cases, organizations will offer users some benefit if they opt-in to using a biometric system. For other organizations, users will have to opt-out of an automatically assumed choice to use biometrics. Finally, some companies may leave the customer with only one choice, to either use biometrics or relinquish the right to use the company's services.

4. Reminders

Financial companies are required to provide customers with an annual reminder of the bank's privacy policy. Companies using and storing biometrics may want to periodically remind customers that their biometrics are on file. The companies may also want to update the data or re-register users to minimize the effects of change on the biometric over time. Despite providing a "something-you-are" authentication factor, biometrics do change over time. A National Institute of Standards and Technology study found that:

> Accuracy dropped as subject age at time of capture increased, especially for subjects over 50 years of age. This effect may be due largely to image quality, which is known to vary by age.[39]

Reminders to users that they have biometric data on file, and that it can (or should) be updated periodically, can help to maintain a company's ability to use the biometrics as required.

5. Opportunities

At a minimum, any process that the company creates around registering and using biometrics should allow opportunities to identify and fix erroneous data. For example, customers should be given the ability to make address changes, or to alert the company that the voice on record with the voice identification system is not truly the customer's voice.

6. Repeated Consents

For companies using biometrics in simple, obvious ways, consent might be given once, and repeated use by consumers would serve as sufficient reminder that the company has and uses biometric information about the customer. In other cases, companies may make sophisticated, complex systems built on biometric data collected long ago or far away. For these companies, it may make sense to periodically ask customers if they still accept the company's storage and use of their biometric information.

7. Extension

Occasionally, new opportunities to use existing biometric data will arise. Or companies may find that they want to use an additional piece of biometric data. In cases where a company changes its policy on the collection, use, and storage of biometric data, it should at least notify the customer of such a change. It may also wish to re-apply principles of notification and consent.

G. Other Issues

The context in which the customer has given his consent matters. If a customer clicks on an "I agree" box on a Web site containing a lot of text, only some of which is germane to biometric data collection, the customer may not really have consented. Similarly, if a customer reads a screen on a kiosk-type system telling him to sign up for biometrics by placing his thumb on a fingerprint reader at a particular time during the process, and the customer places his thumb on the reader at the right time, then customer consent has probably been achieved.

In general, a customer should actively participate in a multistep process by selecting options and by reading a clear and easy-to-understand explanation of each process step, including the overall effect of the process.

The first question users of biometric systems ask is always some variation of "Is it safe?" In asking such a question, the user is seeking to clarify the consequences of the theft of his biometric data. Obviously, the intent is somewhat different for facial recognition systems, voice recognition systems, and other non-fingerprint systems. Nonetheless, the question is a good one. Biometric systems are not well understood or widely used yet.

The most widespread system of biometrics is law enforcement's collection of fingerprints, front facial and profile photos, though such system has only recently moved into the computer space. This system, because it collects a full fingerprint, could theoretically be used to create a fake fingerprint. This fake fingerprint, whether attached to a real or fake finger, might be used to fool a fingerprint system.

Besides law enforcement, most biometric systems try to use and store specific measurements, not a full version of the body part or parts being measured. A handprint-geometry reader measures the length and thickness of each finger and hand; it then stores the measurements. Thumb, forefinger, middle, ring, and pinky fingers are each measured and compared to expected values. These measurements are made at several points along the finger's length, and the specific places that are measured vary among different manufacturers.

For each faceprint, retina, gait, etc., different manufacturers have determined their own individual system of measurements and comparisons. Therefore, a fingerprint taken by one system might have no value for any other system. However, manufacturers do even more to make storage of biometrics safe.

First, as we've described, most systems take a set of measurements and store information related to the measurements, rather than a full copy of a fingerprint, face, or other body part. These measurements are called minutiae and are a mathematical and measurement-based representation of a person's face, retina, or gait. Minutiae are similar to the mathematical and cryptographic concept of a one-way function. These functions are interesting because they are easy to calculate going forward, but are nearly impossible to reverse. It is easy to turn a fingerprint into minutiae, but virtually impossible to re-create a fingerprint from minutiae. A math-based example of this concept would be the raising of a number to the power of 4. It is easy to calculate 5^4 or 5×5×5×5, but it is much harder to calculate the fourth root of 625. Thus, the storage of minutiae in most biometric systems should be considered safe with respect to the risk of the re-creation of the fingerprint based on minutiae stolen from a biometric database.[40]

Within the system, during the initial registration of the user, a sample set of minutiae is collected and tied to a particular name or identity. Somewhere in the system, the identity is given a set of privileges. Such privileges might include entry into a tanning bed, access to a top-secret weapons facility, or faster entry into the airport. They could even include the ability to log onto a computer system.

After registration, the user's biometric is collected again during each use of the service. At these times, the biometric would be compared to the entire database for identification of the user. For greater security, the system may seek to authenticate the user. In authentication, the system uses a log-in name, a user's badge, or something else to pull up the specific user's records, and then compares the biometric just taken to the specific biometric rather than just look for any biometric that matches.

Beyond storing only minutiae, most vendors incorporate additional security measures into their systems. As a secondary measure, most vendors encrypt data between the collection point and the main data storage system. This prevents attackers from intercepting the data and also prevents an attacker from simply replaying a previous good fingerprint. Encryption also prevents an attacker from inserting minutiae stolen from a database. The stolen data, when inserted into the connection between the collection point and the central system, would be decrypted again by the central system, effectively making it unreadable to the central system and incomparable to other minutiae in the system.

Third, as mentioned, each vendor has its own measurement and comparison algorithm; in fact, the comparison algorithm is one of the areas in which vendors often compete. Each vendor designs its system to be the fastest and most accurate, capable of producing no false positives[41] and few false negatives.[42] Naturally no system is perfect, and scientists and engineers discuss the crossover point of a system. The crossover is when the system switches between the fewest false positives and the fewest false negatives. This underlying difference in algorithms among biometric vendors limits how much damage a stolen set of minutiae could create.

Besides these three measures, which are mostly intended to improve the security of the system and which provide security for the user's minutiae, some vendors and system implementers go further. Systems may be located locally within a building, preventing an attacker physical access. Well-designed biometric systems for access control to physical spaces would protect themselves as much as the rest of the space. Many systems will limit or prevent Internet access.

Some systems will provide the user with a smart card to store his fingerprint, making the user responsible for the security of his stored minutiae, while still making the data available to authenticate the user to the system. Organizations may use some other electronic token that stores user's biometric data in a distributed fashion, rather than store such data centrally. In all of these cases, users may feel more secure because the organization does not retain their data.

> ### Recommendation Five
>
> Companies should have a data retention and data destruction policy, especially with respect to biometric data, and follow that policy rigorously. Data that a company does not have in storage cannot be used in unintended ways, cannot be subpoenaed, and cannot be lost or stolen. Some have suggested that the most mature companies in identity management will be those that keep just enough information to conduct business, but no more, and for no longer than necessary.[43]

VI. Summary

The unanticipated consequences of pervasive biometrics are likely to include significant invasions of personal privacy as biometric files are aggregated with other information to build meta-database profiles. Private biometric information will be "purchased" by companies offering bargains, discounts or cash in exchange for a customer's data. Public biometric information will continue to be gathered by law enforcement and businesses until some of those databases are used to harm people, and the public backlash will likely lead to significant legal restrictions on capture and use of public biometric data. In addition, a delayed public debate will ensue about what biometric data may be legally linked with behavior traits to exclude certain people or to offer benefits to others. De facto standards will emerge from government for collection, use, and storage of a universal biometric identifier. In order to deal with these unanticipated consequences, companies should follow the recommendations set forth in this chapter.

Notes

1. *Passwords Revealed By Sweet Deal*, BBC NEWS, http://news.bbc.co.uk/1/hi/technology/3639679.stm (last visited April 3, 2006).

2. Security Focus, *Company Requires RFID Injection*, Feb. 10, 2006, http://www.securityfocus.com/brief/134 (related story about implanted RFID devices for authentication).

3. Check Clearing For the 21st Century Act, Oct. 28, 2004, 12 U.S.C. §§ 5001-5018, with accompanying regulations at 12 C.F.R. part 229.

4. Mary Snow, *No Smoking*, CNN, Jan. 26, 2005, http://www.cnn.com/2005/US/01/26/no.smoking.

5. David Wyld, *Biometrics at the Disney Gates*, Secure ID News, March 2, 2006, http://www.secureidnews.com/library/2006/03/02/biometrics-at-the-disney-gates/.

6. The European Parliament and Council have implemented a directive "on the protection of individuals with regard to the processing of personal data and on the free movement of such data." Directive 95/46/EC, 1995 O.J. (L 281), 31-50 (EC). This directive, which calls for stringent conditions on the collection and use of an individual's private information, serves as a basis for the privacy laws of the member European countries.

7. *Id.* art. 6(1).

8. *Id.* art. 17.

9. The U.S. Congress has acted to protect financial information (Gramm-Leach-Bliley Financial Services Modernization Act of 1999, 15 U.S.C. §§ 6801-6809), protect health care information (Health Insurance Portability and Accountability Act (HIPAA), Standards of Privacy for Individually Identifiable Health Care Data, 45 C.F.R. pts. 160, 164, August 14, 2002), and protect information relating to children (Children's Online Privacy Protection Act of 1998 (COPPA), 15 U.S.C. §§ 6501 *et seq.*).

10. "The Constitution does not explicitly mention any right of privacy. In a line of decisions, however, going back perhaps as far as *Union Pacific R.R. Co. v. Botsford*, 141 U.S. 250, 251 (1891), the Court has recognized that a right of personal privacy, or a guarantee of certain areas or zones of privacy, does exist under the Constitution." Roe v. Wade, 410 U.S. 113, 152 (1973) (citing Stanley v. Georgia, 394 U.S. 557, 564 (1969); Terry v. Ohio, 392 U.S. 1, 8-9 (1968); Katz v. United States, 389 U.S. 347, 350 (1967); Boyd v. United States, 116 U.S. 616 (1886); Olmstead v. United States, 277 U.S. 438, 478 (1928) (Brandeis, J., dissenting); Griswold v. Connecticut, 381 U.S. 479, 484-485 (1965); Meyer v. Nebraska, 262 U.S. 390, 399 (1923).

11. *Id.*

12. *Supra* note 10.

13. *See, e.g., In the Matter of* Microsoft Corp., F.T.C. No. 012-3240 (2002) (where the FTC accused Microsoft of misrepresenting its security and privacy measures taken with regard to the Passport service). *See also* F.T.C. v. Eli Lilly & Co., F.T.C. No. 012-3214 (2001) (where FTC accused Eli Lilly of breaching its own claims of privacy and security with regard to a Prozac e-mail list when the company released e-mail addresses of list members to other list members).

14. The Gramm-Leach-Bliley Act may restrict the dissemination of any personally identifying information, but only as it relates to the financial information of that person. The act does not restrict the capture and dissemination of a biometric indicator attached to the name of the person leaving that indicator.

15. *See, e.g.,* Allright Parking Sys., Inc. v. Deniger, 508 S.W.2d 127 (Tex. Civ. App. 1974) (holding bailee liable for theft of bailor's property); Robertson v. Clark Bros. Builders, Inc., 786 S.W.2d 602 (Mo. Ct. App. S.D. 1990) (same); McPherson v. Belnap, 830 P.2d 302 (Utah Ct. App. 1992) (same).

16. *See, e.g.,* Warren St. John, *Today's Bank Robber Might Look Like a Neighbor,* N.Y. Times, July 4, 2004, *available at* http://www.nytimes.com/2004/07/03/national/ 03ROB.html?ei=5090&en=f5493916d50c9377&ex=1246593600&partner=rssuserland& pagewanted=print&position; Steven Levy & Brad Stone, *Grand Theft Identity,* Newsweek, July 4, 2005, *available at* http://www.msnbc.msn.com/id/8359692/site/ newsweek/print/1/displaymode/1098/.

17. The largest overhaul in federal government treatment of financial services since the great Depression of the 1930s came in the package of the Gramm-Leach-Bliley Financial Services Modernization Act of 1999 (*supra* note 10), followed by accompanying regulations throughout the early 2000s. By the time this law and its regulations were enacted, the banking and securities industries had been heavily invested in information technology for more than 15 years with networking computer models dominating business computing through the 1990s.

18. 45 C.F.R. pts. 160, 162, 164.

19. *See* COPPA, *supra* note 10.

20. *See, e.g.,* Complaint, Texas v. Sony BMG Music Entm't, LLC (Travis County District Ct. filed Nov. 21, 2005); Settlement Agreement, *In re* Doubleclick, Inc., Aug. 26, 2002, www.oag.state.ny.us/press/2002/aug/aug26a_02_attach.pdf (a privacy-related settlement agreement between Doubleclick and the attorneys general of Arizona, California, Connecticut, Massachusetts, Michigan, New Jersey, New Mexico, New York, Vermont, and Washington).

21. Cal. Civ. Code §§ 1798.29, 1798.82.

22. At least 22 other states have followed California's lead and have passed security breach notification laws, including New York (A4254, A3492), Texas (SB 122), Washington (SB 6043), Ohio (HB 104), North Carolina (SB 1048), Maine (LS 1671), Louisiana (SB 205), Montana (HB 732), New Jersey (A4001/S1914), and Nevada (SB 347). A full list of states that have implemented notification laws and the relevant statutory text are available at http://www.pirg.org/consumer/credit/statelaws.htm (last accessed March 27, 2006).

23. Jonathan Krim, *Banking Rules Address Theft of Customers' Private Data,* Wash. Post, March 24, 2005, at E1, *available at* http://pqasb.pqarchiver.com/washingtonpost/ access/811735901.html?dids=811735901:811735901&FMT=ABS&FMTS= ABS: FT&fmac=&date=Mar+24%2C+2005&author=Jonathan+Krim&desc=Banking+ Rules +Address+Theft+Of+Customers%27+Private+Data

24. *See, e.g.,* U.S. Const. amend. IV.

25. *See, e.g.,* The Privacy Act of 1974, 5 U.S.C. § 552(a); The Electronic Communications Privacy Act, 18 U.S.C. § 2510.

26. Examples of California state bills concerning the collection or use of biometric data that failed to become law include AB 2193 (Biometric and Personal Information Act) (introduced on Feb. 23, 2000); SB 169 (introduced Feb. 5, 2001, and amended on March 15, 2001, to regulate biometrics); AB 46 (introduced Dec. 2, 2002); AB 215 (introduced Jan. 29, 2003); SB 682 (Identity Information Protection Act of 2005) (shelved by the Assembly's Appropriations Committee in August 2005). The text of these bills is *available at* http://www.leginfo.ca.gov/bilinfo.html.

27. John Crewdson, *Internet Blows CIA Cover,* Chicago Tribune, March 12, 2006, *available at* http://www.chicagotribune.com/news/nationworld/chi-060311ciamain-story,1,123362.story?ctrack=1&cset=true.

28. Robert Lemos, *ChoicePoint Data Loss May Be Higher Than Reported*, March 10, 2005, http://news.com.com/ChoicePoint+data+loss+may+be+higher+than+reported/2100-1029_3-5609253.html

29. *See* 18 U.S.C. § 2710.

30. *See* Press Release, International Civil Aviation Organization, Biometrics Featured at Global Symposium on Machine-Readable Travel Documents (Aug. 18, 2005), http://www.icao.int/mrtd/download/documents/Biometrics%20deployment%20of%20Machine%20Readable%20Travel%20Documents% 202004.pdf

31. Associated Press, *FBI Apologizes to Lawyer Held in Madrid Bombings,* May 25, 2004, http://www.msnbc.msn.com/id/5053007/.

32. Shaila Dewan, *F.B.I. Apologizes for Failing to Identify Murder Suspect*, N.Y. TIMES, May 5, 2005, *available at* http://select.nytimes.com/gst/abstract.html?res=F00610FD3C540C768CDDAC0894DD404482t.

33. Electronic Privacy Information Center, http://www.epic.org/privacy/biometrics/ (last visited April 2, 2006).

34. Better Business Bureau, http://www.bbbonline.org/privacy/threshold.asp (last visited April 2, 2006).

35. Direct Marketing Association, http://www.the-dma.org/privacy/creating.shtml (last visited April 2, 2006).

36. Electronic Privacy Information Center, http://www.epic.org/privacy/ (last visited April 2, 2006).

37. Spyware and adware are generic names for malicious software that collects user information surreptitiously, slows or ties up the computer's resources, distributes collected data without the user's knowledge and understanding, and, in some cases, delivers pop-up or other unwanted advertising.

38. Fingerprint Vendor Technology Evaluation, http://fpvte.nist.gov/report/ir_7123_summary.pdf (last visited April 2, 2006).

39. It is actually much easier to steal a person's fingerprint by following him to a restaurant or bar. Most people use the utensils, drink from glasses, and otherwise leave their fingerprints all over the restaurant without trying. Any episode of any police television show demonstrates the ability to capture fingerprints easily. In 2002, a Japanese researcher demonstrated how to use these latent fingerprints to make fake fingers that were accepted by the fingerprint readers he was testing. *Fun with Fingerprint Readers*, Crypto-Gram Newsletter, May 15, 2002, http://www.schneier.com/crypto-gram-0205.html#4).

40. False positives occur when the system incorrectly identifies a match between someone trying to get into the system and a valid user of the system.

41. False negatives occur when the system incorrectly prevents a valid user from using the system.

42. High West, http://radio.weblogs.com/0146815/2005/08/02.html (Aug. 2, 2005, 21:40 EST).

Index